Chemical Bonding

Mark J. Winter
Department of Chemistry, University of Sheffield

Series sponsor: **ZENECA**

ZENECA is a major international company active in four main areas of business: Pharmaceuticals, Agrochemicals and Seeds, Specialty Chemicals, and Biological Products.

ZENECA's skill and innovative ideas in organic chemistry and bioscience create products and services which improve the world's health, nutrition, environment, and quality of life.

ZENECA is committed to the support of education in chemistry and chemical engineering.

OXFORD
UNIVERSITY PRESS

OXFORD
UNIVERSITY PRESS

Great Clarendon Street, Oxford OX2 6DP
Oxford University Press is a department of the University of Oxford.
It furthers the University's objective of excellence in research, scholarship,
and education by publishing worldwide in

Oxford New York

Athens Auckland Bangkok Bogotá Buenos Aires Calcutta
Cape Town Chennai Dar es Salaam Delhi Florence Hong Kong Istanbul
Karachi Kuala Lumpur Madrid Melbourne Mexico City Mumbai
Nairobi Paris São Paolo Singapore Taipei Tokyo Toronto Warsaw

with associated companies in Berlin Ibadan

Oxford is a registered trade mark of Oxford University Press
in the UK and in certain other countries

Published in the United States
by Oxford University Press Inc., New York

First published 1994
Reprinted 1995, 1996, 1997, 1998, 2000

A catalogue record for this book is available from the British Library

Library of Congress Cataloging in Publication Data
(Data available)

ISBN 0 19 855694 2

Printed in Great Britain by
The Bath Press, Bath

Series Editor's Foreword

Understanding of the basis of periodicity and the nature of chemical bonding is fundamental to inorganic chemistry. All of our discussions of structure, physical properties, and chemical reactivity stem from these basics.

Oxford Chemistry Primers are designed to give a concise introduction to all chemistry students by providing material that would usually be covered in an 8–10 lecture course. As well as providing up-to-date information, this series will provide explanations and rationales that form the framework of an understanding of inorganic chemistry. Mark Winter explains the fundamentals of chemical bonding in a clear and readable way. This particular primer is aimed at first-year university students and should provide them with a sound base to underpin later inorganic chemistry courses.

John Evans
Department of Chemistry, University of Southampton

Preface

This short book is intended to introduce some concepts of chemical bonding in a descriptive and essentially non-mathematical fashion. The coverage is not intended to be rigorous and is aimed at the student rather than the lecturer. It is therefore hoped that this short text will find a place *alongside* textbooks containing more detailed formal coverage and mathematical descriptions. A complementary discussion of the bonding in transition metal systems is included in a further title in this Primer series.

The electron density diagrams were calculated with the HyperCard™ program Hyperorbitals. I am very grateful to many people for their help during the preparation of this book, especially to Duncan Bruce, Alison Cherry, Bill Clegg, Mike Morris, and Barry Pickup for their comments and corrections of some of my misconceptions. I am also grateful to Douglas W. Franco of the Instituto de Fisica e Quimica de São Carlos-USP, Brazil, at where a large part of the manuscript was prepared.

Sheffield M.J.W.
January, 1993

Contents

1 Atomic structure

Chemistry is about *molecules*. Molecules are *atoms* chemically bound together. To understand chemical bonding, it is clearly necessary to appreciate something of the nature of atoms, the participants in bonding. It is not the purpose of this book to examine in any detail historical aspects of the evolution of atomic theory, but a historical preamble is justified.

1.1 Early views of atomic structure

More than 2000 years ago, Democritus, a Greek philosopher, asked what would happen if one were to divide a piece of matter into two, and into two again, and again, and so on. Would there come a point at which it would not be possible to divide that piece of matter into two? Or could one continue this process forever? Democritus felt that there *was* a point at which further divisions would not be possible. He called the resulting 'fundamental' bits of matter *atoms*. He felt the differences between different kinds of matter were due to differences in the size, shape, kind, and proportions of the constituent atoms. A remarkable deduction, but one that was not then generally accepted. The problem with his theory was that he had no means of verification.

Democritus: 470–380 BC

Lucretius was a Roman poet who wrote an epic, *De Rerum Natura* in which his ideas of nature were based upon the ideas of Democritus, as modified by Epicurus. He saw a primal universe containing an infinite number of atoms swirling around. These met to form small particles, which combined to make larger particles, and these combined finally to form worlds. He postulated the chance interaction of atoms as building up living creatures, whose soul was a material aggregate of atoms, but more subtle than the body. These ideas seem quite modern, but were not accepted at the time.

Lucretius: *c*.94–*c*.55 BC

De Rerum Natura: On the Nature of Things

Empedocles, and later Aristotle, had very different ideas. They believed that earthly substances are a composite of *matter* and *essence*. The substance was supposed to consist of *prime matter* whose only property is that it is material. Superimposed upon this prime matter was a combination of four *elements*: earth, air, fire, and water. These elements were not material, but qualities rather like solidity, volatility, energy, and liquidity. The proportions of these four elements were reputed to make a particular kind of material. Aristotle was a great philosopher and his thoughts concerning classification in biology are profound, but his ideas on chemistry are less useful. However, his influence was immense and his ideas on almost any area of science were generally regarded as truth until the seventeenth century. It was more than 2000 years before the Democritean concept of atoms was seriously revived.

Aristotle: 384–322 BC

Boyle addressed the composition of matter in *The Sceptical Chymist* in 1661. In the 1800s, John Dalton's experiments showed that matter consisted of elementary particles, or *atoms*. These atoms combine in fixed relative

Robert Boyle: AD1627–91

proportions to form *molecules*. However, the nature of the atoms was far from clear.

Joseph J. Thomson: 1856–1940

Around the turn of the century, it was appreciated that atoms consist of electrically charged particles, both positive and negative. The physicist J.J. Thomson proposed a model for the structure of the atom based upon a relatively large, positively charged, amorphous mass in which much smaller, negatively charged, electrons are embedded.

1.2 Rutherford's experiments

Ernest R. Rutherford: 1871–1937

Some of the earliest sensible conclusions concerning atomic structure were based on experiments carried out by Rutherford around 1906, and a little later (1909) by Geiger and Marsden. These experiments involved directing a stream of α-particles (helium nuclei) towards a piece of gold foil (Fig. 1.1). Radium, encased in a lead block as shown, provided a parallel beam of α-particles.

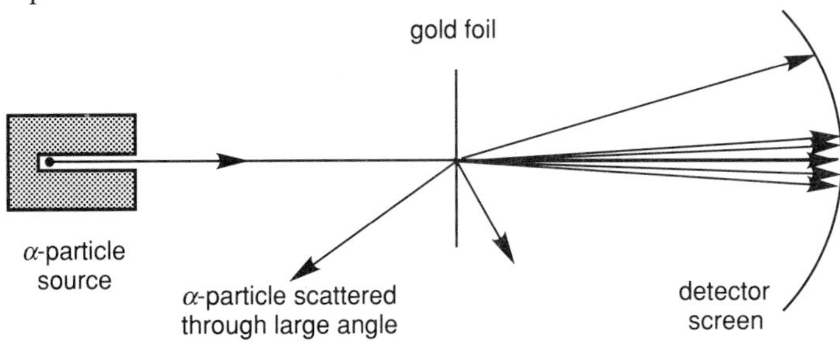

Fig. 1.1 Rutherford's apparatus for an analysis of α-particle scattering by gold foil.

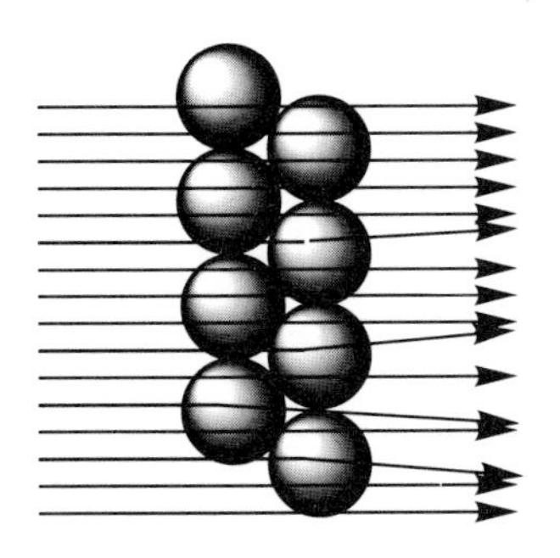

Fig. 1.2 The expected result of the Rutherford α-particle scattering experiment, assuming the Thomson model.

It was anticipated, based upon the Thomson atomic model, that the stream of α-particles would pass straight through the foil (Fig. 1.2), with perhaps just a few particles being slightly deflected. This is because the Thomson model postulates a rather uniform distribution of matter throughout the atom.

The results obtained were a great surprise. Certainly, most of the α-particles *did* pass straight through the foil. However, a few were deflected through quite large angles, sometimes greater than 90°. Such large deflections would not occur if the Thomson model was reasonably correct. The experimental results led Rutherford to suggest that an atom consists of a very small, but relatively massive, positively charged nucleus, surrounded by electrons (Fig. 1.3). The components of the atom are held together by electrostatic forces. Virtually all the mass of the atom is associated with the positively charged nucleus, even though it is so small relative to the size of an atom. The close approach of an α-particle to the nucleus of a Rutherford atom occasionally results in a large scattering angle. The diameter of the nucleus is now known to be about $1/10^5$ the size of the atom itself.

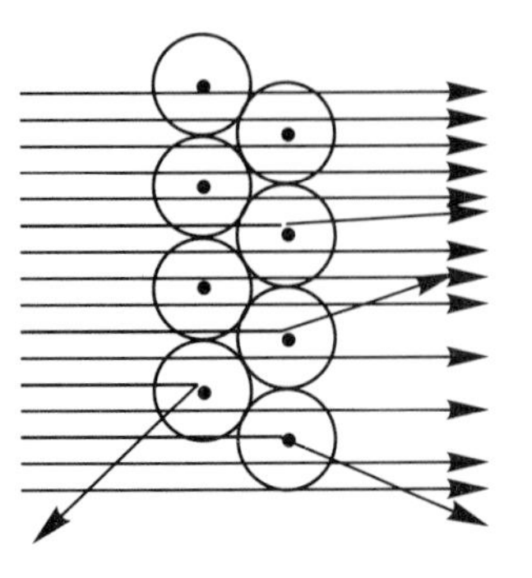

Fig. 1.3 The results of the Rutherford α-particle scattering experiment.

Rutherford suggested that the reason most α-particles pass through gold foil with no or little deviation is simply that the atom is largely empty space and just doesn't interact much with the intruding α-particle. However, just occasionally, one of the α-particles collides with, or comes very close to, the 'massive' nucleus. On these occasions, the resulting interaction is sufficient to cause a large deviation in the path of the α-particle, even to the extent of deflecting the α-particle back along the direction from which it came. Clearly these ideas portray an atom very different from the Thomson model.

1.3 The Bohr model of the atom

Niels Bohr: 1885–1962

The simplest sensible model for atomic structure based on Rutherford's conclusions is the *Bohr model*, a model put forward by the Danish physicist Niels Bohr in 1913. This model allowed him to successfully account for the emission and absorption spectra of atomic hydrogen.

He suggested that electrons (negatively charged) *revolve* around the nucleus (positively charged) at certain fixed distances in a set of *orbits*. His hydrogen atom (Fig. 1.4) consists of a small central nucleus of mass m_n, around which one small negatively charged electron of mass m_e 'orbits'. The orbital velocity of the electron is v and the distance at which the electron orbits is r.

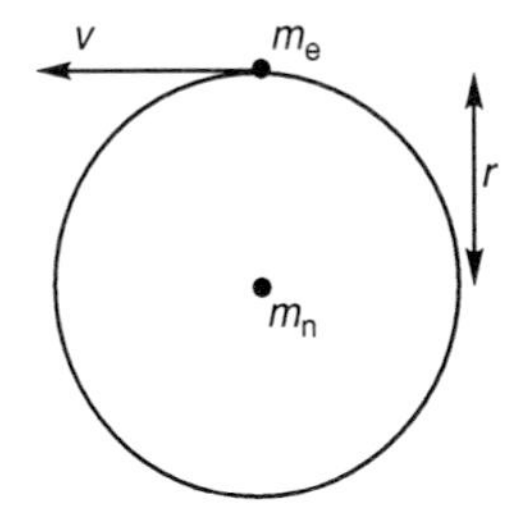

Fig. 1.4 Bohr's hydrogen atom.

Left to itself, at any instant, the hydrogen atom electron would proceed in a straight line along the path denoted by the velocity v. However, the electron is *not* left to itself. It is negatively charged, while the nucleus is positively charged. There is therefore an *electrostatic* attraction between them. The effect of this attraction is to prevent the electron moving away from the nucleus along the straight line defined by v and instead the electron is constrained to orbit the nucleus in a circular orbit. The effect of this is that the *magnitude* of v remains constant, but the *direction* of v constantly changes to track the tangent of the circular orbit.

$E = KE + PE$

It would be useful to know the energy associated with the hydrogen atom electron in its Bohr orbit. The total energy, E, of the electron is the sum of the *kinetic energy*, KE, and the *potential energy*, PE.

If the charge on the hydrogen nucleus is $+e$ and the charge on the single electron is precisely $-e$, then it is relatively straightforward to show that one *only* needs to know the distance r to calculate E. Equation 1.1 expresses this. Mathematically, the energy, E, is a *function* of r, that is $E = f(r)$.

$$E = -\frac{e^2}{4\pi\varepsilon_0 r} \tag{1.1}$$

ε: Greek letter epsilon
π: Greek letter pi

In Eqn 1.1, e is the value of *electron charge* (-1.60218×10^{-19} C), ε_0 is the *permittivity of vacuum* (8.85419×10^{-12} J^{-1} C^2 m^{-1}), and $\pi = 3.14159$. At this point, the Bohr model for the hydrogen atom has problems. An electron circulating about a nucleus will produce a magnetic field and electromagnetic interactions will cause energy loss, resulting in the electron spiralling into the nucleus! While this is happening, it should continuously radiate energy. A second problem is that all values of r are apparently allowed. Since the atomic spectra (Section 1.4) of hydrogen consist of discrete lines and is not continuous, this cannot be the case.

Bohr therefore placed some *constraints* upon his atomic model to stop this happening. These constraints are central to his model. He suggested that only certain distances r for orbits are allowed. It is appropriate to call the orbits with these special values of r Bohr orbits. Since only *certain* orbits are allowed, the radii of the Bohr orbits are said to be *quantized*. Each of the Bohr orbits has a particular energy associated with it, and these energies can

quantum, *n.* quantity; amount; a naturally occurring fixed minimum amount of some entity which is such that all other amounts of that entity occurring in physical processes in nature are integral multiples thereof (Phys.); *pl.* quanta.

quantize: *v. (t.)* 1. (Phys.); to restrict (a physical quantity) to one of a set of fixed values. 2. (Math.); to limit to values that are multiples of a basic unit.

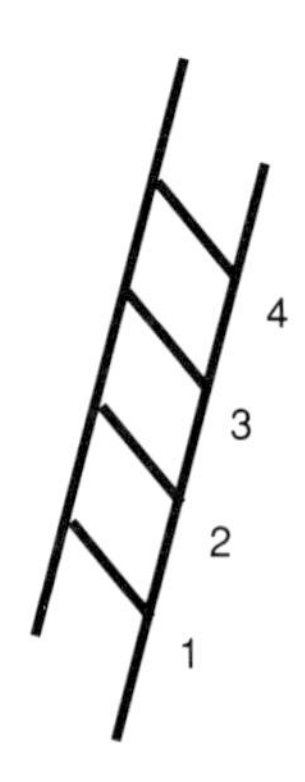

Fig. 1.5 Ladder quantum numbers.

be calculated using Eqn 1.1. Since the orbits are quantized, the energies of the various orbits are also quantized.

Quantization is a concept that many find mysterious but it need not be that difficult to understand. Imagine standing on a ladder (Fig. 1.5). Without divine powers, your position on the ladder can only be on a rung. You cannot stand between rungs. Now label the rungs 1, 2, 3, etc. from the bottom. These labels are *quantum numbers* which describe your position on the ladder. If the ladder quantum number is 3, you are standing on the third rung.

Usually, the Bohr hydrogen atom exists with the electron rotating in just one orbit, the Bohr orbital with the *smallest* allowed value of r. When discussing chemical bonding involving hydrogen it is only normally necessary to consider that one orbit. However, there are circumstances when the Bohr hydrogen atom exists with the electron in some other orbit with a different value for r. Such circumstances are particularly important when discussing spectroscopy, as will also become apparent shortly.

The Bohr orbits are labelled by their quantum numbers. The orbit closest to the nucleus is labelled with the quantum number 1, the next is labelled 2, and so on. One of the beauties of these quantum numbers is that a simple equation (Eqn 1.2) gives the values of r based upon those quantum numbers, which for the Bohr orbits are given the symbol n.

$$r = \text{constant multiplied by } n^2 = k \times n^2 \qquad (1.2)$$

For the first Bohr orbit, $n = 1$, and so its value of r is simply 1^2 multiplied by the constant k. For the second Bohr orbit, $n = 2$, and so its value of r is 2^2 multiplied by the constant k. Provided one knows the value of the constant, one can generate values for r based upon the quantum number n. While beyond the scope of this text, it is relatively straightforward to show that the value of the constant is given by Eqn 1.3. The quantity h is *Planck's constant* (6.62608×10^{-34} J s) and m_e is the *electron mass when stationary* (9.10939×10^{-31} kg).

$$k = \frac{4\varepsilon_0 h^2}{\pi m_e e^2} = 52.918 \text{ pm} \qquad (1.3)$$

1 atomic unit = 52.918 pm = $1a_0$

1 pm = 10^{-12} m

$1a_0$ = 1 *Bohr*

The constant k corresponds to the length 52.918 pm. This length frequently crops up in theoretical chemistry. It is also a very handy size when discussing objects with atomic dimensions. It is therefore quite common to use this length as an atomic unit of size, rather as one uses kilometres or miles to describe distances between cities. This value is given its own symbol, a_0, and its own name, the *Bohr*. The first values of the Bohr orbit radii r as calculated using Eqn 1.2 are given in Table 1.1. The radii of the Bohr orbits increase in size according to a square law (n^2), with values of $1a_0$, $4a_0$, $9a_0$, $16a_0$, etc. (Fig. 1.6).

Since the radii r of the Bohr orbits are quantized, it follows naturally that only *certain* energies exist for that atom: the *energy* of the atom is also quantized. Equation 1.1 relates the energy E to the orbit radius. Substituting the value of r and the constant given by Eqns 1.2 and 1.3 gives Eqn 1.4. This looks somewhat fearsome until one recognizes that all the terms other

Table 1.1. Radii and energies of the first four Bohr radii

n	n^2	r/pm	E_n/J	E_n/cm^{-1}
1	1	52.918	-21.799×10^{-5}	−109679
2	4	211.671	-5.450×10^{-5}	−27420
3	9	476.259	-2.422×10^{-5}	−12186
4	16	846.683	-1.362×10^{-5}	−6855

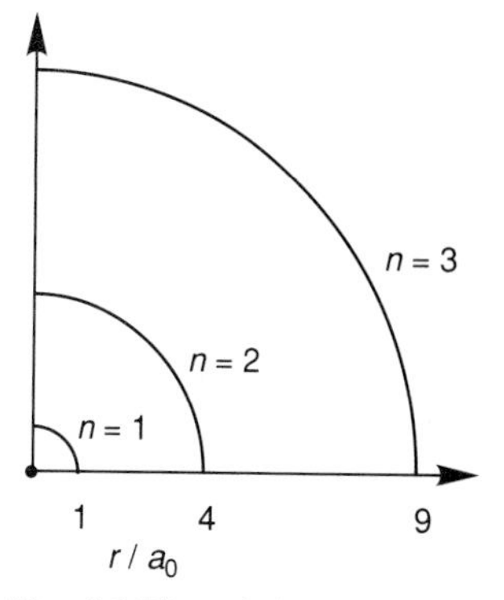

Fig. 1.6 The relative sizes of the first three Bohr radii.

than n are constants. They can therefore be gathered together into a single constant, call it k'. In Eqn 1.4, the energy is written as E_n which is a shorthand to indicate that the energy E is a function of (that is, determined by) *only* the quantum number n.

$$E_n = -\frac{m_e e^4}{8\varepsilon_0 n^2 h^2} = -\frac{k'}{n^2} \qquad (1.4)$$

∞ means *infinity*

Orbits closer to the nucleus correspond to values of E which are most negative. The value of E for $n = \infty$ (E_∞) is set, by definition, to be zero, a reference point. Therefore as n gets larger, the value of E_n gets closer to zero, but in smaller increments ($1/n^2$), exactly the opposite behaviour to the radii of the Bohr orbits which increase in size according to a square law (n^2).

Each of the energies calculated for the Bohr orbits is referred to as an *energy level*. The energies of the energy levels corresponding to each Bohr orbit can be plotted on an *energy level diagram*. These are a very common device in chemistry and it is a good idea to be familiar with them.

Each energy level is labelled with a number, 1, 2, 3, etc. These are the same *quantum numbers* introduced earlier. These particular quantum numbers are referred to as the *principal quantum numbers*. In effect, these quantum numbers are *names* for the energy levels. Unlike the rungs on a set of ladders, there is no need for equal spacing of the energy levels. Figure 1.7 shows a system of *energy levels*. Each energy level refers to the energy of an electron contained in an orbit. In atoms, there is an infinite number of possible levels. Their energies converge to a limit, labelled by ∞ and the energy $E_\infty = 0$. Most higher energy levels occur at energies rather close to zero.

1.4 Spectroscopic properties of hydrogen atoms

Any decent model of atomic structure must lead to a successful representation of chemical bonding. It must also explain the spectroscopic properties of the atoms of that element. In this regard the Bohr model of the hydrogen atom is rather successful.

Light from a standard light bulb is more or less continuous. That is, light from that bulb passed through a prism and projected on to a screen gives a full or *continuous* spectrum showing all wavelengths. However atoms absorb light, and if the light from the bulb is passed through an atomic gas before refraction through the prism, the resulting light when projected onto a screen is *not* continuous. The *absorption spectrum* shows a series of narrow dark lines where specific colours are missing. The patterns of dark lines are characteristic to each element and the technique for their analyses is called

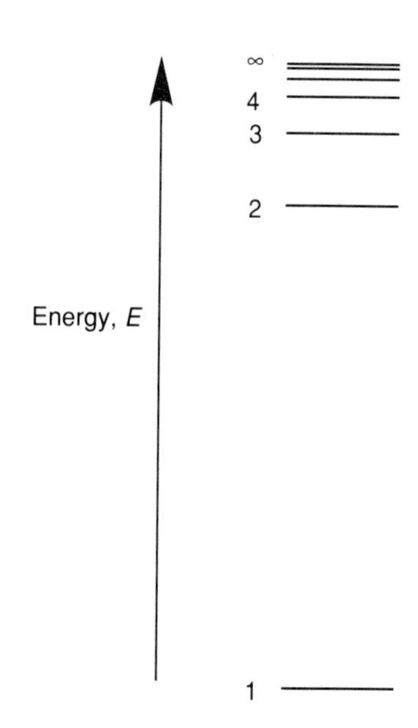

Fig. 1.7 A system of energy levels.

atomic absorption spectroscopy. The atomic absorption spectrum of hydrogen is a consequence of movement of the electron from the *ground state energy level* to some *higher energy level*. When an electron is promoted from the ground state level to a higher energy level in an atom, then the atom gains the difference in energy between the two levels from the light source. The energy of the light photon absorbed is given by a famous equation which relates the energy difference, ΔE, to the frequency, ν of the radiation.

Note the distinction between v (velocity) and ν (Greek letter nu, frequency).

$$\Delta E = h\nu \qquad (1.5)$$

When atoms are heated to high enough temperatures, perhaps by an electric discharge, they *emit* light. The colours are characteristic; hydrogen is red, sodium is orange, and so on. The emitted light is *not continuous*, that is, not all frequencies are emitted. In fact, the spectrum consists of a dark background upon which appear series of bright *emission lines*. Observation of such spectra is achieved by use of *atomic emission spectroscopy*. In atomic emission spectroscopy, the energy source promotes the ground state electron to a higher energy level. The excited electron then moves back down to the ground state. When an electron passes from a high energy level to a low energy level in an atom, then that atom emits a quantum of light, a photon. The energy of that photon is equal to the difference in energy between higher and lower energy levels. It is normal to express the position of a line in the atomic spectrum in *wavenumbers*, $\bar{\nu}$, the reciprocal of the wavelength of the light (Eqn 1.6). It is clear that the quantum number for the electron changes as it moves between the various energy levels. Since frequency, ν, is related to wavelength, λ (Fig. 1.8), by the relation $\lambda\nu = c$ (where c is the speed of light = 2.9979×10^8 m s^{-1}) then substitution gives Eqn 1.7.

$$\bar{\nu} = \frac{1}{\lambda} \qquad (1.6)$$

$$\Delta E = h\bar{\nu}c \qquad (1.7)$$

This equation shows why the wavenumber is so useful. Since c and h are constants, the wavenumber is directly proportional to the energy. Double the wavenumber, double the energy.

For hydrogen, the absorption spectrum consists of one series of lines. These appear as a consequence of promotion of the hydrogen 1s electron (n = 1) to higher energy levels (n = 2, 3, 4, etc.). The position of the first line gives the difference in energy between the n = 1 and the n = 2 level. This is the smallest jump that the electron is able to make in the ground state atom on absorption of energy, and so this transition is at the longest wavelength (lowest frequency) in the series of lines. The next highest frequency line corresponds to the 1→3 transition. Since the energy levels get closer to each other as n increases (Fig 1.7), the difference in energy between the 1→x and 1→(x + 1) transitions gets less as x increases. In fact, as x tends towards ∞, the lines converge to a *limit* denoted by the transition 1→∞. This is an interesting transition since it represents the minimum amount of energy required to remove the electron completely from the hydrogen atom. This

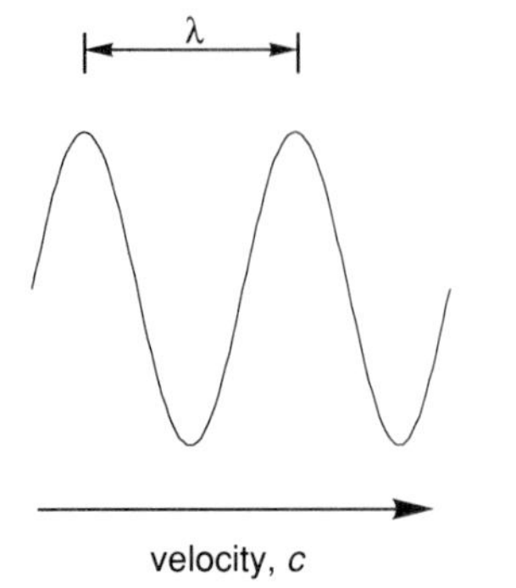

Fig. 1.8 Definition of wavelength, λ, for light.

quantity of energy corresponds to the *ionization energy* of hydrogen. One way to determine the ionization energy of hydrogen is therefore to record the absorption spectrum of atomic hydrogen.

Ionization energy: the energy required to remove an electron from the highest occupied energy level to that corresponding to $n = \infty$.

The emission spectrum of hydrogen is more complex and consists of several series of lines. The electron moves up the energy levels of hydrogen on absorption of energy from an external energy source, perhaps an electric discharge. Excitation means that the electron acquires more energy. The electron remains for a short time in the higher energy level until it loses that energy by moving down to a lower energy level. This need not be the ground state energy level. The electron may move down in a series of jumps until it eventually reaches the ground state. As the electron jumps down to a lower energy level, the excess energy is shed by the *emission* of a quantum of light. So, sodium atoms in a Bunsen flame are excited to higher levels by the flame and fall to lower levels with the *emission* of light. For sodium the strongest emission line is in the orange region of the spectrum, accounting for the characteristic yellow-orange colour of the sodium flame.

Figure 1.9 shows several ways in which the hydrogen electron could move to lower energy levels after promotion to a higher level. For convenience, they are grouped together into sets. When the electron moves back to, say, the ground state, a photon is emitted whose energy is equal to the drop in energy experienced by the electron on moving down to the ground state from the higher energy level.

The set of transitions shown on the left all terminate at the $n = 1$ level. One way to represent these transitions is $2 \rightarrow 1$, $3 \rightarrow 1$, $4 \rightarrow 1$, and so on. Since the energy levels converge to a *limit* which corresponds to the energy level for $n = \infty$, then the series of emission lines converges to a limit at the $\infty \rightarrow 1$ transition. This set of transitions down to the ground level corresponds exactly in terms of wavenumber positions to the set of transitions shown in the *absorption* spectrum of atomic hydrogen.

A second set of lines (centre) is a set of transitions which terminate at the $n = 2$ level can be represented as $3 \rightarrow 2$, $4 \rightarrow 2$, $5 \rightarrow 2$, etc. Yet another set of lines terminate at the $n = 3$ level. Each of these sets of transitions terminating at a particular lower level shows up as a series of converging

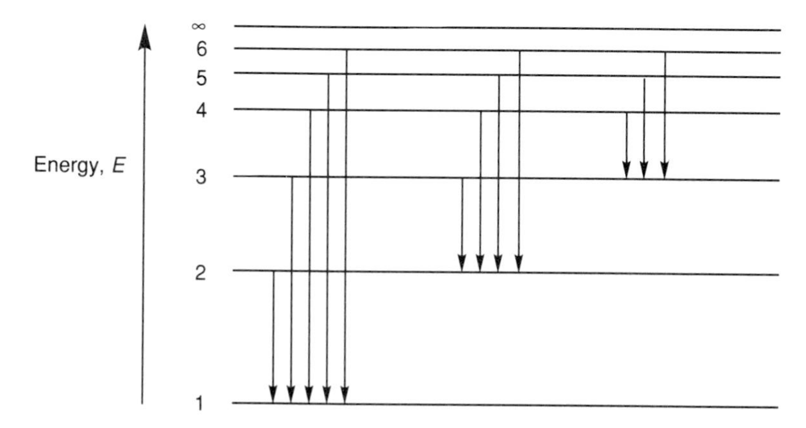

Fig. 1.9 Transitions between energy levels in the hydrogen atom (not to scale).

Table 1.2. Named emission series in the atomic spectrum of hydrogen

Discoverer	Transition
Lyman	$n' \rightarrow 1$
Balmer	$n' \rightarrow 2$
Paschen	$n' \rightarrow 3$
Brackett	$n' \rightarrow 4$
Pfund	$n' \rightarrow 5$

lines in the emission spectrum. Other sets of lines correspond to other sets of transitions, all of which correspond to various $n' \rightarrow n$ transfers. Each of the various series of lines is named after its discoverer (Table 1.2). Sets of lines for $n' \rightarrow 6$ and higher exist, but are not named after anyone. The Bohr model neatly accounts for the atomic spectra of hydrogen. The sets of transitions arise as a result of transfer between the various allowed Bohr orbits, and the positions of the spectroscopic lines are correctly predicted by very simple equations based upon the Bohr quantum numbers. The positions of each line in each set of lines is given by the Rydberg equation (Eqn 1.8).

$$\text{Frequency} = R\left(\frac{1}{n^2} - \frac{1}{n'^2}\right) \tag{1.8}$$

$$\text{Frequency} = 109678.764\left(\frac{1}{2^2} - \frac{1}{n'^2}\right) \tag{1.9}$$

$$\text{Frequency} = 109678.764\left(\frac{1}{2^2} - \frac{1}{5^2}\right) = 26322.9\ \text{cm}^{-1} \tag{1.10}$$

$R = 109\ 678.764\ \text{cm}^{-1}$

In this equation, n is the quantum number expressing the energy level *to which* the transition proceeds while n' is the energy level *from which* the transition starts. The letter R is a number called the *Rydberg constant*. It is a requirement that $n' > n$. The values for n' in each series are integral up to infinity, ∞. Thus each line of, say, the *second* series of lines is given by Eqn 1.10. To calculate the value of the $5 \rightarrow 2$ transition, write 5 as the value for n', into Eqn 1.9 to give Eqn 1.10.

The Rydberg constant R is related to the constants of Eqn 1.4 and the correspondence of the predicted value of R to the experimentally determined value of R was a triumph for the Bohr model of the hydrogen atom. The idea of an electron circling the nucleus is simple and the concept therefore gained widespread acceptance. However, it soon became clear that the Bohr model for hydrogen did not tell the whole story. For instance, the atomic spectrum of atoms in a magnetic field is different from that when there is no magnetic field. This is the *Zeeman* effect. The simple Bohr model does not account for this. However, the Bohr model was rescued temporarily by Arnold Sommerfeld, who modified the Bohr model. He accounted for the Zeeman effect by suggesting that there might be elliptical orbits in atoms in addition to spherical orbits. A description of this requires a *second* quantum number.

The Bohr–Sommerfield model is good for hydrogen but breaks down for heavier elements. For instance, it does not satisfactorily account for their atomic spectra. Neither does it account for the periodic properties of elements expressed in the format of the standard periodic table (inside back cover). The Bohr model eventually gave way to a more sophisticated model based on *wave mechanics*.

1.5 Particle or wave?

Electrons and other fundamental atomic particles are often best not regarded as little hard lumps of matter (like billiard-balls) at all. Electrons are rather tenuous entities whose properties include those traditionally associated with *both* particles and waves. It is possible to *diffract* beams of electrons through crystals in a similar way to the diffraction of light by a set of regularly spaced lines or X-rays by crystals. This is a property generally perceived as being associated with *waves* rather than *particles*. It is also possible to *weigh* electrons, traditionally a particle property. In a related fashion, light, which one might assume to have just simple wave properties, has properties associated with particles, such as an effective momentum, and it comes in quantized 'packets' called *photons*. Mathematically, whether to regard electrons or light as particles or waves is very much determined by the problem under discussion. Sometimes, one representation is more useful and appropriate than the other. It is well to remember that whichever the style of mathematics, the result is a description of the *properties* of electrons or light, not the nature of electrons or light.

The 'particle–wave duality' was expressed by de Broglie. His postulate was that all matter possesses characteristics of both waves and particles. He expressed this duality in a famous and profound equation (Eqn 1.11). In this equation, the wavelength is expressed as a function of mass, m, and velocity, v (or momentum since momentum = mv).

$$\lambda = \frac{h}{mv} \tag{1.11}$$

This leads to the suggestion that particle and wave phenomena simply relate to different attributes of all matter. This is a confusing idea. Everyday experience suggests that things are either particles or waves but not both. A ball is a particle and sound is a wave. In practice, the so-called duality only becomes significant for very, very small entities such as atomic particles or light. The difficulty in understanding their nature is because it is difficult to relate our everyday experience of light and matter to the apparently conflicting natures of such small particles. Perhaps it seems ridiculous to suggest that 'everyday' matter can have wave properties, but it isn't. The problem lies in the magnitude of the wavelength. For any visible object, the size of the wavelength is so small as not to be perceived. The *effective* wavelength of a particle is only important when the particle is very small.

Consider a cricket ball, mass 5.25 oz (150 g) moving at 100 mph (45 m s^{-1}). Substituting into Eqn 1.11 gives the wavelength of the associated wave as $6.63 \times 10^{-34}/(0.15 \times 45) \cong 10^{-34}$ m, which is *very* small. For comparison, the diameter of an atom is of the order of 10^{-10} m. On the other hand, an electron whose mass when stationary is about 10^{-30} kg moving at the same speed has an associated de Broglie wavelength of about 1.5×10^{-5} m, a much more experimentally observable length.

1.6 From orbits to orbitals

In the Bohr model, electrons orbit the nucleus. In the wave mechanical model the concept of 'planetary' electrons moving around a nuclear 'sun' is less helpful. Two pieces of important information about an electron in an atom are the position and momentum. However, *it is not possible to precisely*

Momentum = mass x velocity
Both *speed* and *direction* are required to define the velocity.

Werner Heisenberg: 1901–1976

determine the position and momentum of an electron at any particular instant. This is an aspect of the *Heisenberg Uncertainty Principle*. This is illustrated as follows. If the position of an electron is to be defined, it is necessary to *look* at it to see where it is. Since it so small, some kind of supermicroscope is required. If an object is to be seen, it is necessary for the illuminating light to possess a wavelength smaller than the size of the object. The size of an atom is far smaller than the magnitude of visible light wavelengths, therefore it is necessary to use very short wavelength light smaller than that of an atom (if the position of the electron is to be defined within the atom). Such wavelengths are associated only with very energetic electromagnetic radiation (since $E = h\nu$). Light of such short wavelengths is not visible to the eye. Electromagnetic radiation is light, composed of *photons*. Photons possess particle-like properties, including an effective momentum. Therefore, if light is used to illuminate the electrons so that they are rendered visible, the effect is that the illuminating photons hit the electrons and transfer some of their momentum to the electron. This means the original momentum of the electron is changed, and it is not possible without further irradiation to look again at the electron to determine this resulting change, an act which would *again* change the momentum, and so on

Looking at an object, in this case an electron, *changes its properties* (momentum and position). An unobserved object necessarily possesses different properties to an observed object.

If it is desired to make more accurate measurements of the position, the shortest wavelengths possible of illuminating radiation must be used. However, this means even more energetic light, and the consequence is that on illumination, even more of a momentum change affects the electron. So a price is paid for increased precision in determining the position of the electron, in that the momentum of the electron changes even more. A rigorous analysis of this and related phenomena shows that the more precisely the *position* of the electron is known, the less precisely the *momentum* is known, and vice versa. Further, there is a definite limit to the precision with which the momentum and position can be determined. This is expressed by the Eqn 1.12 which relates the uncertainties in determining position and momentum to Planck's constant, h.

Planck's constant: 6.62608×10^{-34} J s

$$\text{Uncertainty in position} \times \text{uncertainty in momentum} \geq \frac{h}{4\pi} \qquad (1.12)$$

The fact that it is not possible to precisely determine the position and velocity of a particle is not disastrous. Instead, a different mathematical treatment to analyse the properties of the electron is used. For atoms, it is usual to talk about the *probabilities* of an electron being in a particular place at some time. Although the position of the electron cannot be defined exactly, the *probability* of finding the electron at some particular point, at some particular time is a number that *can* be calculated. If the probability of finding an electron at some point is high, then the *electron density* is said to be high at that point. The *probability* of finding the electron at a particular point equates with the term *electron density* for that point.

The *electron density* or *probability density* representations of electrons in atoms describe an electron as located within a specific region of space with a particular electron density at each point in space. This space and a description of its contained electron density is called an *orbital*, a term clearly borrowed from the Bohr model.

A good way to visualize the situation for the hydrogen atom in the ground state is as follows. Imagine sliding an amazingly thin piece of 'photographic' film right through the centre of the atom, so that the hydrogen nucleus lies in the plane of the film. The photographic film has the property that every time the electron particle passes through it, a dot on the film results, and the electron passes along its path without hindrance. The next stage is to wait for a while and so record many dots on the film. The result of this experiment might look something like that in Fig. 1.10. This represents a cross-section of the orbital. The result is a cloud of dots and this is a representation of electron density. The density of dots in any particular region is a pictorial representation of the probability density or electron density in that region. At first sight the result looks surprising. It is immediately clear that the electron does *not* keep to a constant distance from the nucleus. In fact, the density of the dots increases towards the nucleus, and is at a maximum at the nucleus.

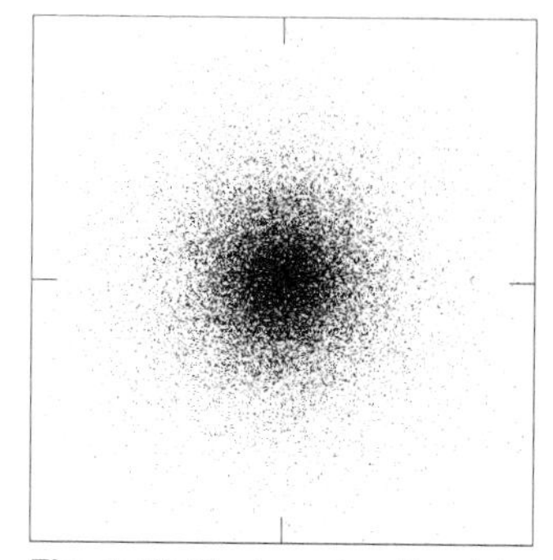

Fig. 1.10 Electron density plot (25 000 dots) for the hydrogen atom in the ground state. The square is 400 pm wide, with the nucleus at the centre.

The 'dot density' representation is a useful graphic representation, but such diagrams are only conveniently and accurately generated with a suitable computer program. Other representations are possible, and easier for the individual to draw on a piece of paper. One common way to proceed is as follows: given a mathematical formula to express the electron density, it is possible to calculate a surface at which the electron density is *equal* over all the surface and *within which* there is a, say, 40% chance of finding the electron. Another equal value surface corresponds to, say, an 80% chance of finding the electron. This surface of equal density possesses a shape and is the basis of the *contour representation*. The contours represent boundaries within which there is some specified chance of locating the electron. The bigger the volume, the greater the chance of finding the electron. Again this is quite a good way of representing the orbital, but its a pain drawing all those circles, and if many accurate contours are to be drawn, again a suitable computer program is required.

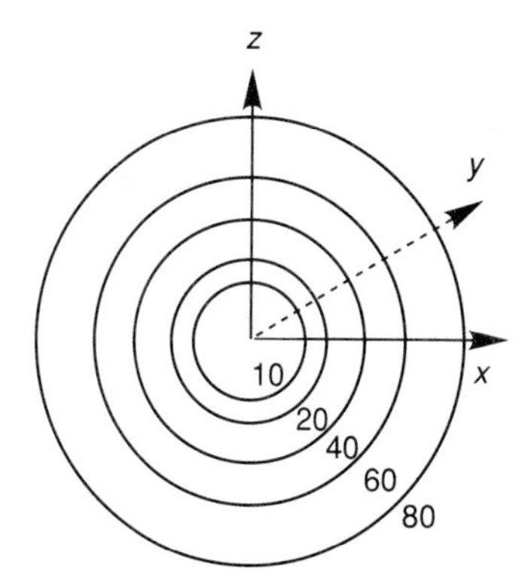

Fig. 1.11 Contour representation of electron density for the hydrogen atom in the ground state (values in %).

It is often adequate to draw just *one* contour, and to use that one contour simply to denote the *shape* of the orbital rather than to make any attempt to indicate electron density. Diagrams of this type are often referred to as *boundary representations*. The boundary representation is often the most convenient way to represent an orbital since it is the quickest to draw, but of course it doesn't tell much about the structure of the orbital, other than a general impression of its shape. Frequently the percentage contour line being used is not even labelled—it is often just the *shape* of the orbital that is of interest. If necessary, the single contour or boundary could be defined as that surface within which, say, 90 or 95% of the total electron density is contained. This very simple representation will be used frequently throughout this book and is in common usage throughout the chemistry community.

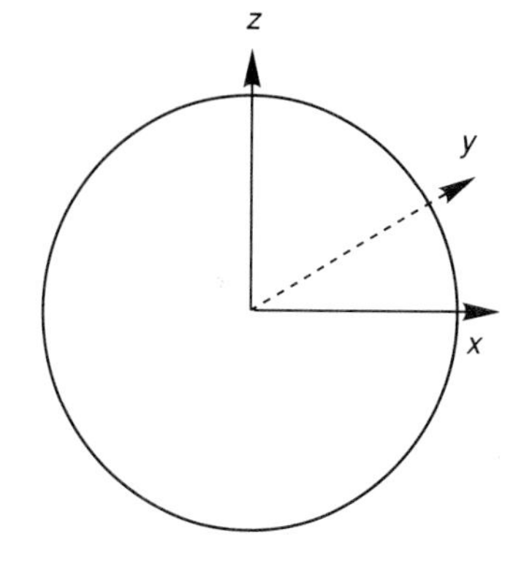

Fig. 1.12 Boundary representation of hydrogen in the ground state.

1.7 The quantization of orbitals and energies

It is a requirement of the Bohr–Sommerfeld model that the orbits are quantized, but the quantization is introduced, apparently arbitrarily, to make the facts fit. The Bohr–Sommerfield model does not explain *why* there is quantization of the energy levels. It is the wave-like properties of electrons that provide the clue. Schrödinger proposed that the way forward was to emphasize the wave nature of electrons and to produce equations to describe the wave-like properties of electrons within atoms. He used a method called *wave mechanics* to describe the nature of the electrons in their energy levels (orbitals) using wave equations.

Erwin Schrödinger: 1887–1961

The behaviour of other, much more familiar systems are satisfactorily described by wave equations. A string on a stringed musical instrument produces only notes of particular wavelengths, the fundamental and some harmonics. That is, the vibrations of a string fixed at both ends are *quantized*. This is a perfectly natural consequence of the *ends* of the string being *fixed* in position. The fixing of the ends of the strings constitute '*boundary conditions*' for the wave equations used to describe the notes and it is this imposition of boundary conditions for motions of the string which makes the wavelengths produced by the string quantized. The wave equations used to describe the behaviour of the hydrogen atom electron similarly require the imposition of perfectly natural boundary conditions.

The Schrödinger wave equations have caused problems for generations of students and will continue to do so, but the mathematics will not be examined here. However the results of the solutions to these equations will be used. In most cases it is not possible to solve the Schrödinger wave equation exactly, but it is possible to do so for a one-electron atom such as hydrogen. There is more than one solution. In fact there are many. The solutions to these wave mechanical differential equations do one thing quite naturally, and that is to introduce quantization. This is because of the imposed boundary conditions. The boundary conditions are that the equation must be *finite*, *single-valued*, and *continuous*. Since there is a 100% chance of finding the electron somewhere in the orbital, it is clear that the equation cannot go off to infinity at any point: all values of the equation must be *finite*. There can only be one probability value of finding the electron at any one point: the equation must be *single-valued* at all points. It is unreasonable for the equations to introduce sudden jumps in value on moving from one point to another: the equation must be *continuous*.

Each solution corresponds to one energy level and each energy level in the hydrogen atom is predicted correctly by the solutions to the wave equations. Solutions to such equations often involve *families of equations* whose specification requires quantum numbers. In the case of the Schrödinger equation for hydrogen, the solutions require *three* quantum numbers to specify each. These numbers are represented by n, l, and m_l (Table 1.3).

Application of these rules allows a table of energy levels to be built up. There is just one energy level for $n = 1$, four for $n = 2$, and nine for $n = 3$. Each set of orbitals corresponding to any one value of n is termed a *shell*. There is therefore one orbital in the first shell, four in the second, and nine in the third. Each set of orbitals within a shell for which m_l is the same is

Table 1.3. The first three orbital quantum numbers

Label	Description
n	*principal quantum number* or more graphically, the *orbital size quantum number.* It can take any integral value from 1 to infinity (∞).
l	*azimuthal* or *angular momentum quantum number,* or more graphically, the *orbital shape quantum number.* For any particular value of n, l can adopt any integer value from 0 to (n–1).
m_l	*magnetic quantum number,* or more graphically the *orbital orientation quantum number.* Allowed values depend on the value of l and can take any integer value from $-l$ through zero to $+l$.

Table 1.4. Relationship of quantum numbers to orbital names

n	1	2	2	2	2	3	3	3	3	3	3	3	3	3
l	0	0	1	1	1	0	1	1	1	2	2	2	2	2
m_l	0	0	-1	0	1	0	-1	0	1	-2	-1	0	1	2
name	1*s*	2*s*	2*p*	2*p*	2*p*	3*s*	3*p*	3*p*	3*p*	3*d*	3*d*	3*d*	3*d*	3*d*

termed a *subshell*. There is therefore only ever one orbital in each subshell for which $l = 0$, three orbitals in each subshell for which $l = 1$, and five orbitals in each subshell for which $l = 2$. There is just one subshell in the $n = 1$ shell, two in the $n = 2$ shell, and three in the $n = 3$ shell.

As for the Bohr hydrogen atom, it is possible to calculate the energy associated with each wave function solutions for the hydrogen atom. For the hydrogen atom, all the orbitals within a particular shell are *degenerate*, that is their energies are *precisely* equal. For instance, in the $n = 3$ level the nine orbitals are degenerate, This is only true for one electron systems such as hydrogen and is not true for any system containing more than one electron.

1.8 Naming orbitals

Orbitals are named after the quantum numbers. The first part of the name is simply the principal quantum number. The second is linked to the *orbital shape quantum number*. The origin of the names *s*, *p*, *d*, and *f* is historical and connected with the nature of atomic spectroscopy lines. Orbitals with higher values of l (>3) are given alphabetic names starting from g. Conversely it should be clear that the name of a particular orbital defines the quantum numbers for that orbital.

The single solution for which $n = 3$ and $l = 0$ is called the 3*s* orbital. The three solutions with $n = 3$ and $l = 1$ are all called 3*p* orbitals. Each of these three has a different value of m_l. The five solutions with $n = 3$ and $l = 2$ are all called 3*d* orbitals (Table 1.4). Each of these has a different value of m_l.

Table 1.5. Dependence of orbital name on l

l	name	origin of name
0	*s*	sharp
1	*p*	principal
2	*d*	diffuse
3	*f*	fundamental

1.9 The hydrogen 1*s* orbital

The *lowest* energy solution for the wave function for the hydrogen atom is called the *ground state* and this solution to the wave function is given a

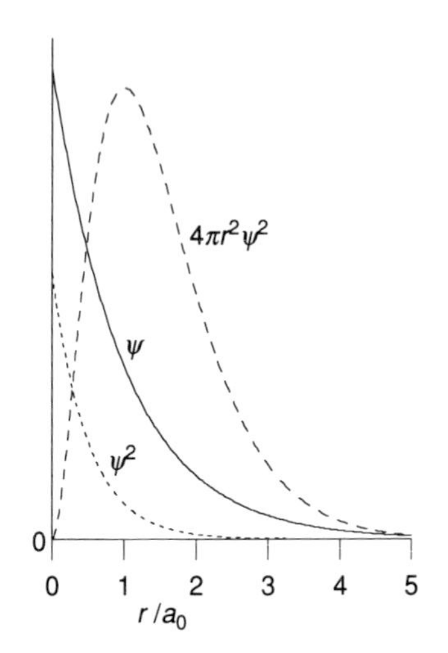

Fig. 1.13 Graph of the wave function ψ_{1s} for the hydrogen 1*s* orbital, together with plots of ψ^2 and the radial distribution function.

name, the 1*s* orbital (since $n = 1$, $l = 0$, and $m_l = 0$, the only possibility for the first shell). The wave equation for the 1*s* orbital, ψ_{1s}, [pronounced (p)sī 1*s*] is given by an exponential function, Eqn 1.13, in which the only variable is *r*, the distance from the nucleus. The value a_0 is the Bohr radius (52.9 pm). The number e is approximately 2.71828 and is a number that, in its own way, is as significant as π. It is not to be confused with another *e*, the charge on an electron. For many purposes in chemistry, the equations look simpler using the Bohr radius, a_0, as the *unit of length*, rather than picometres or Ångstroms. Using a_0 as unit of length reduces Eqn 1.13 to Eqn 1.14.

$$\psi_{1s} = \frac{1}{\sqrt{\pi}}\left(\frac{1}{a_0}\right)^{3/2} e^{-r/a_0} \quad (1.13)$$

$$\psi_{1s} = \frac{1}{\sqrt{\pi}} e^{-r} \quad (1.14)$$

The *maximum* value for ψ_{1s} occurs at the nucleus ($r = 0$, see Fig. 1.13). The value for ψ_{1s} depends *only* on the distance away from the nucleus and *not* on direction. The symmetry of the ψ_{1s} wave function is *spherical*. The value of ψ_{1s} is positive for all values of *r* and non-zero out to *r* = infinity.

$\psi^2 \equiv$ *electron density*

It is now appropriate to relate the electron density discussed earlier to the wave function. It is largely useless to relate values of the wave function itself to any physical property. This is because there is no direct physical significance! However the value of ψ^2 *does* have a physical significance, since ψ^2 *corresponds to electron density* The value of ψ^2 is the probability of finding an electron at that point or *electron density* at that point. A plot of ψ^2 for the 1*s* orbital is superimposed upon that of ψ in Fig. 1.13.

The electron density is greatest in the vicinity of the nucleus. This is *not* the same as saying the electron is most likely to be found at the nucleus. After all, there is only one point at $r = 0$, while there are many at, say, the surface of the sphere with $r = 100$ pm. Loosely speaking, the number of points associated with a particular value of *r* goes up with the surface area of the sphere, that is with $4\pi r^2$. To get a more useful picture, the *radial distribution function*, a description of how much electron density is found at a given value of *r* is required.

The radial distribution function shows that the electron is more likely to be found at the Bohr radius than at any other radius. Although the electron density ψ^2 is greatest at the nucleus, the chances of finding it there are small since there is so little volume at the nucleus. To understand the radial distribution function better it is appropriate to consider some apparently unrelated concepts. Consider a box filled with a gas of some density. The total weight of gas in the box is given by volume × density (Fig. 1.14).

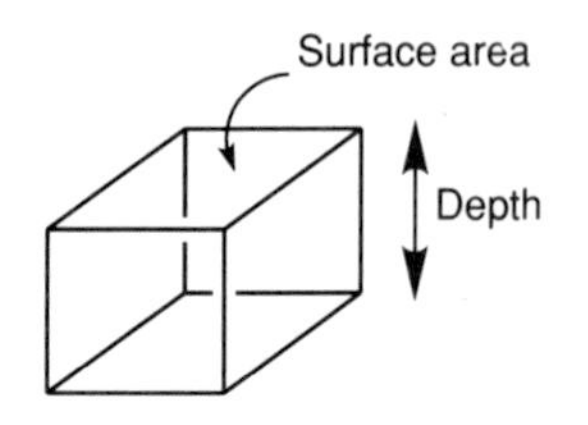

Fig 1.14 The volume of a box.

The volume is given by the lid surface area multiplied by the depth, or thickness, of the box. If the gas density is constant, the calculation is trivial. The situation for electron density is analogous, but more difficult to calculate since the electron density is *not* constant and the shape of the orbital is spherical, at least in the case of hydrogen in the ground state. Mathematically, the way to proceed is through an *integration* of electron density over the volume of the orbital. It is appropriate here just to visualize the integration process. This is conveniently done with recourse to an onion. Mentally, one can unravel a spherical orbital into shells (not to be confused with shells of orbitals!) in just the same way that an idealized onion can be unravelled into shells. The interesting property of each unravelled spherical 'orbital onion shell' is that if the ring is reasonably thin, then the electron density throughout that 'orbital onion ring' is approximately constant. Therefore if the volume of the 'orbital onion shell' is known, the total amount of electron density within that shell can be calculated using Eqn 1.15. The volume of a *thin* spherical onion shell is also given by the relationship surface area X thickness. The surface area of a sphere is $4\pi r^2$. Since the electron density is ψ^2, and the surface area is $4\pi r^2$, then the total electron density at radius r is given by Eqn 1.16, assuming *thin* shells each of equal thickness.

$$\text{Total in 'orbital onion shell'} = \text{surface area} \times \text{thickness} \times e^- \text{ density} \quad (1.15)$$

$$\text{total} = 4\pi r^2 \psi^2 \times \text{thickness of shell} \quad (1.16)$$

In this case the answer is known. For a hydrogen 1*s* orbital with one electron in it, the sum of the electron density in all the 'orbital onion shells' adds up to one electron. The quantity $4\pi r^2\psi^2$ is proportional to the total amount of electron density at any given value of r. A plot of this function is included in Fig. 1.13. Mathematically, in the integration process, an exactly analogous process is followed, but in which the thickness of the shell is modified so that it becomes vanishingly thin. The radial distribution plot shows that the electron is to be found over a *range* of distances, but that it is *most likely* to be found at the Bohr radius, a_0. Since it is not possible to state precisely where the electron is at any given time, it is better to talk in terms of the *probability* of finding the electron at some given distance.

1.10 Other hydrogen *s* orbitals

The orbital described above is the 1*s* orbital and this orbital contains the electron when the hydrogen atom is in the ground state. As for the 1*s* orbital, the other *s* solutions, 2*s*, 3*s*, 4*s*, etc., are all spherically symmetrical. These solutions represent orbitals that are generally empty, but in principle it is possible for them to become occupied if the ground state electron is *promoted* from the 1*s* orbital to, say, the 2*s* orbital. If the hydrogen atom's electron is in any other orbital than the 1*s*, it is said to be in an *excited state*. An electron in each of the hydrogen 1*s*, 2*s*, and 3*s* orbitals is represented in Fig. 1.15 by dot-diagrams. The scale is the same and the number of dots is the same in each case, allowing a comparison of relative size and of electron densities. A number of important features emerge. The most obvious point is that the orbitals increase in size with increasing principal quantum number. This corresponds to increasing orbit size in the Bohr model of the atom.

The next point concerns *radial nodes*. The 2*s* orbital (Fig. 1.15) shows a ring around the nucleus at which the electron density drops to zero. In fact, in

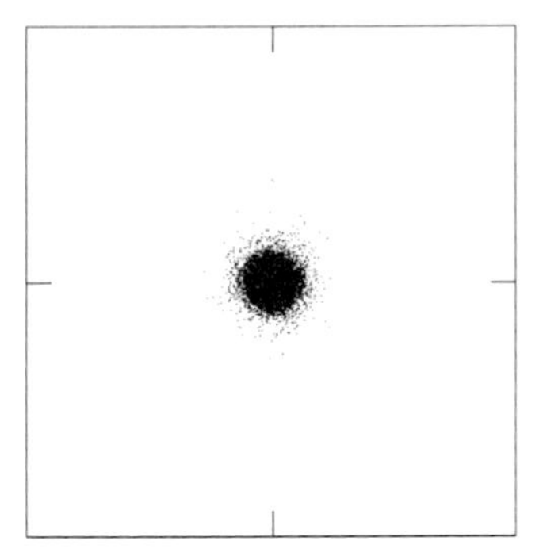

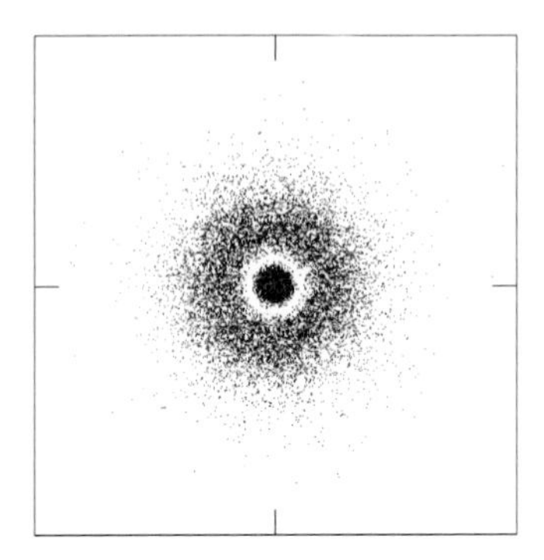

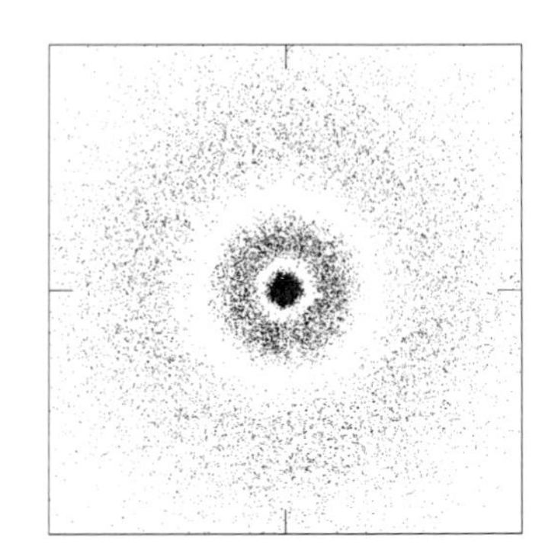

Fig. 1.15 Electron density dot diagrams (15 000 dots) representing the 1*s* (left), 2*s* (centre), and 3*s* (right) orbitals in the hydrogen atom. Each square is 2000 pm wide with the nucleus at the centre of each square.

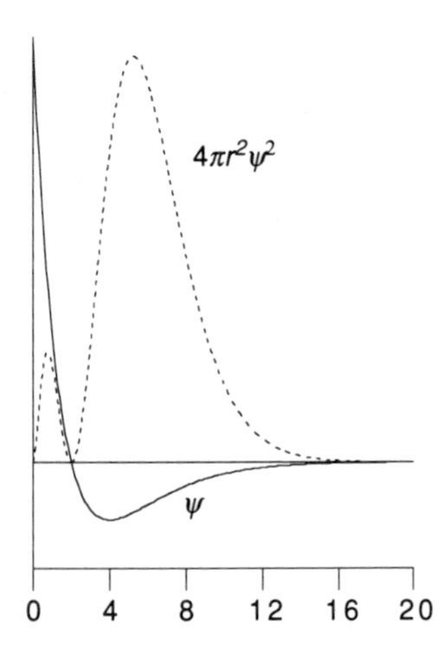

Fig. 1.16 Plot of ψ, and $4\pi r^2\psi^2$ for the hydrogen 2*s* orbital.

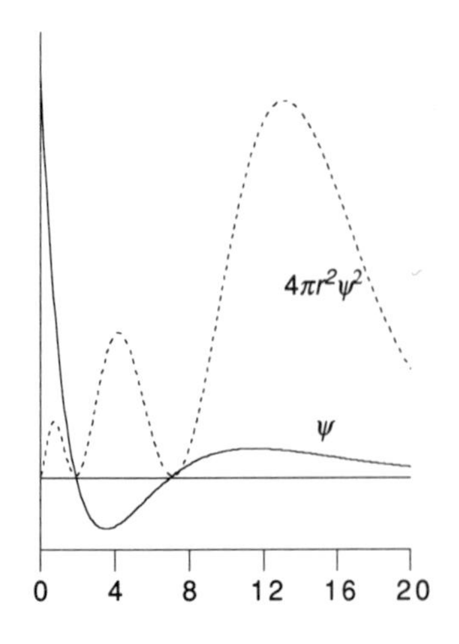

Fig. 1.17 Plot of ψ and $4\pi r^2\psi^2$ for the hydrogen 3*s* orbital.

a three dimensional representation, there is a *spherical* surface at which the electron density drops to *exactly* zero. This feature is referred to as a *node*. The 3*s* orbital has *two* such spherical nodes. In general, the number of radial nodes shown by an *s* orbital with principal quantum number *n* is *n*–1.

The mathematical equations used to represent these orbitals can look somewhat daunting, but they are based upon exponential functions related to that of the 1*s* orbital already seen. Equations 1.17 and 1.18 give the formulae for these orbitals and, for clarity, the unit of length is a_0 as in Eqn 1.13.

$$\psi_{2s} = \frac{1}{4\sqrt{2\pi}}(2-r)e^{-r/2} \tag{1.17}$$

$$\psi_{3s} = \frac{1}{81\sqrt{3\pi}}\left(27-18r+2r^2\right)e^{-r/3} \tag{1.18}$$

It is a worthwhile exercise to plot the wave functions ψ_{2s} and ψ_{3s}. It should then soon be apparent that the radial nodes represent spherical surfaces at which the *sign* of the wave function switches from one sign to the other, that is positive to negative or negative to positive. For instance, the wave function for the 2*s* orbital (Fig. 1.16) changes from being positive to negative at one particular distance, $r = 2a_0$, from the nucleus. The 3*s* orbital wave function (Fig. 1.17) changes sign twice, from positive to negative and back to positive again, which is why there are two radial nodes. The electron density is *not* affected by the *sign* of the wave function. Recall that electron density is calculated from the wave function by squaring it. Squaring a real negative number *always* gives a positive number, that is the electron density is always positive, or perhaps zero if the wave function happens to be zero at that point. This seems sensible since common sense requires that there can never be negative electron density, or a negative probability. Zero or positive, yes, but never negative.

It is instructive to compare the radial distribution plots for the 1*s*, 2*s*, and 3*s* orbitals. These are plotted together in Fig 1.18 on the same scale so that a reasonable comparison can be made. The effective size of the *s* orbitals increases with increasing values of the principal quantum number. It should also be apparent that, for *n* = 2 and *n* = 3, although the electron density is greater in the regions closer to the nucleus, because of the greater volumes

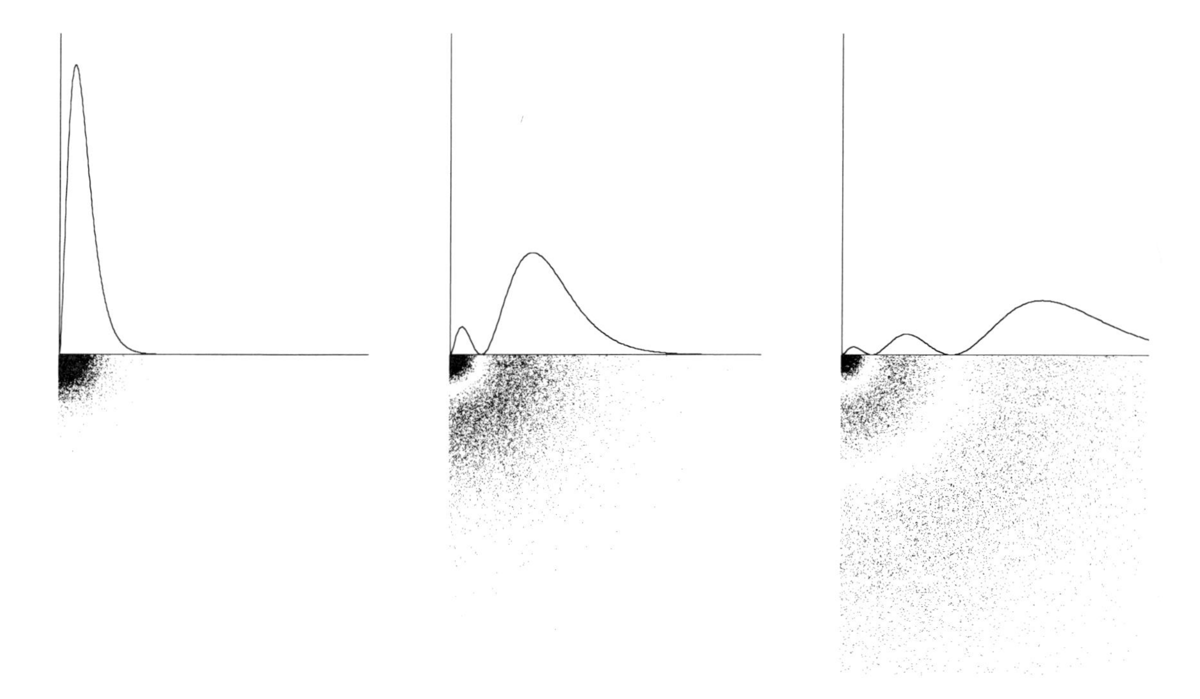

Fig. 1.18 The radial distribution functions for the 1*s*, 2*s*, and 3*s* orbitals related to quadrants of the electron density dot representations of electron density. The 3*s* orbital shows *two* spherical radial nodes, the 2*s* orbital shows *one* , and the 1*s* *none*.

involved further away, the greatest overall amount of electron density is found in the outermost region of the 2*s* and 3*s* orbitals.

1.11 Hydrogen *p* orbitals

The *p* orbitals ($l = 1$) are altogether different in appearance. One of each of the three 2*p* and 3*p* orbitals for the hydrogen atom are plotted in Fig. 1.19 on the same scale used in Fig 1.18. As the principal quantum number increases, the overall size of the orbital increases. However the most striking feature is that the orbitals *are not spherically symmetrical,* that is, they are *directional.* The shape of these orbitals may be generated from these two-dimensional cross-sections by rotating about the vertical axis (straight up and down the page).

Consider first the 2*p* orbital. The top of Fig. 1.19 represents the solution for which $m_l = 0$. Since this orbital has most of its electron density along the *z*-axis, it is called the $2p_z$ orbital. The 2*p* orbitals possess nodes. In each case there is a *nodal plane* dividing the two *lobes* of the orbital (the *xy* plane in the case of the $2p_z$ orbital). As for the *s* orbitals, increasing the principal quantum number is associated with extra nodal surfaces. In the case of the 3*p* orbital (Fig. 1.19, bottom) the second nodal surface is a sphere with radius $6a_0$.

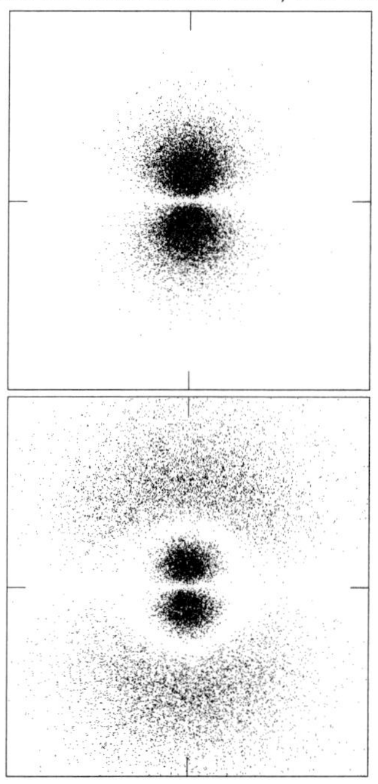

Fig. 1.19 Electron density dot diagrams (15 000 dots) of 2*p* (top) and 3*p* orbitals of the H atom. The squares are 2000 pm wide. The nucleus is at the centre of the squares.

When $n = 2$, the solutions to the wave functions are referred to as the $2p_1$, $2p_0$, and $2p_{-1}$ orbitals (the 1, 0, and −1 subscripts are the three possible values for m_l). For the purposes of this discussion, these three solutions are not totally convenient. The equations of the $2p$ orbitals for which $m_l = +1$ and -1 contain an *imaginary number*, i, the square root of -1. However, mathematically it is perfectly justifiable to 'mix' these two solutions by taking *linear combinations*. The result of taking these linear combinations is two orbitals identical in appearance to the $2p_z$ orbital but which are directed along the x and y axes in the same way as the p_z orbital is directed along the z axis. These two new orbitals are referred to as the $2p_x$ and $2p_y$ orbitals, and their equations (Eqns 1.20 and 1.21) no longer contain i.

$$\psi_{2p_x} = \frac{1}{\sqrt{2}}\left(\psi_{2p_1} + \psi_{2p_{-1}}\right) \tag{1.20}$$

$$\psi_{2p_y} = \frac{1}{\sqrt{2}}\left(\psi_{2p_1} - \psi_{2p_{-1}}\right) \tag{1.21}$$

The three $2p$ orbitals (Fig. 1.20) are identical in shape but differ in the directions in which they point. Therefore it is *not* correct to directly relate the m_l values 1, 0, and -1 to the coordinate axes x, y, and z. For the s orbitals, the wave functions and electron densities are at a *maximum* at the nucleus. For the $2p$ and $3p$ orbitals, the nucleus coincides with nodal surfaces and so the electron density is *zero* at the nucleus.

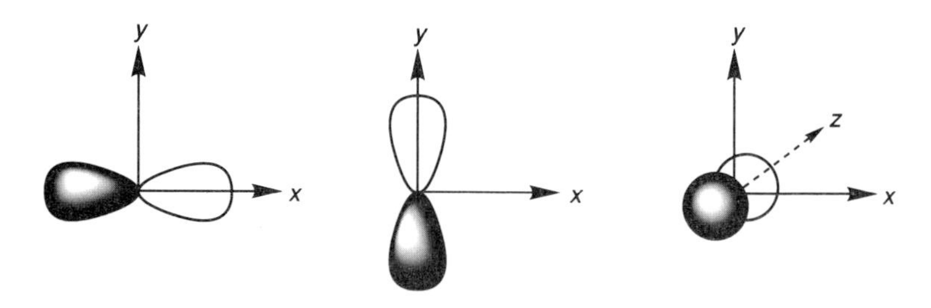

Fig. 1.20 Schematic representations of the $2p_x$, $2p_y$, and $2p_z$ orbitals.

1.12 Wave function shading and signs

Fig. 1.21 The shading indicates the sign of the wave function of a $2p$ orbital.

The schematic representation of a p orbital shwn in Fig. 1.21 contains two pieces of information. The shape of the diagram corresponds to ψ^2, the probability density. The shading indicates the relative sign of the wave function, that is, whether it is positive or negative. The relative signs of the wave function are often explicitly indicated by adding '+' or '–' signs. *These signs have nothing to do with electric charge.* They refer only to whether the value of the wave function is positive or negative. Whether the sign of the wave function is positive or negative is not of particular importance for atoms but is very important when addressing how atoms bond together to give molecules.

1.13 Hydrogen *d* orbitals

There are *five* independent d orbital solutions for each principal quantum number > 2. These correspond to values of m_l of $-2, -1, 0, 1$, and 2. Four of the equations, those for which m_l is not zero, contain i, the imaginary square root of –1 and it is convenient to make linear combinations to produce functions related to the coordinate axes x, y, and z.

Four of the resulting wave functions, $3d_{xy}$, $3d_{xz}$, $3d_{yz}$, and $3d_{x2-y2}$ are similar in appearance (Figs. 1.23 and 1.24), differing only in their relative orientations. The four lobes of electron density in the $3d_{xy}$, $3d_{xz}$, and $3d_{yz}$, orbitals are orientated between the coordinate axes. The $3d_{x2-y2}$ orbital is the same size and shape, but orientated so that the lobes are directed along the x and y axes.

The $3d_{z2}$ orbital (Figs 1.23 and 1.24) is a different shape. One half of the total electron density is located in two major lobes directed along the z axis. The other half of the electron density is located in a ring- or doughnut-shaped region centred around the xy plane. The orbital possesses *cylindrical* symmetry in that a three-dimensional representation is generated by rotating the dot diagram about the z axis. Mathematically, the $3d_{z2}$ orbital function corresponds to a linear combination of $3d_{z2-y2}$ and $3d_{z2-x2}$ functions and an equivalent name for the $3d_{z2}$ orbital is the $3d_{2z2-x2-y2}$ orbital.

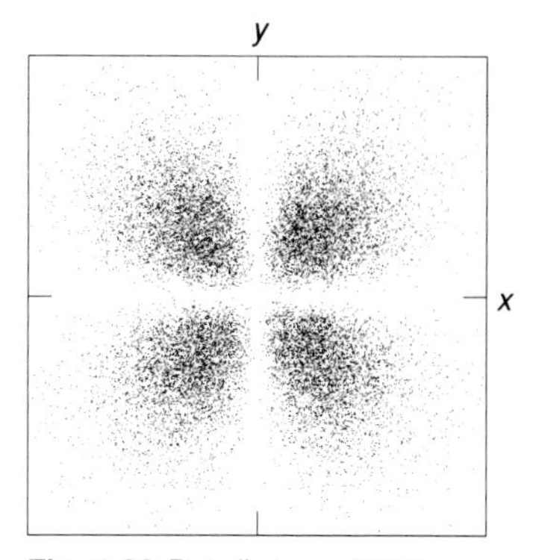

Fig. 1.22 Dot diagram (2000 pm square, 15 000 dots) for the hydrogen $3d_{xy}$ orbital. The yz and xz planes are nodal surfaces. The nucleus is at the centre of the square.

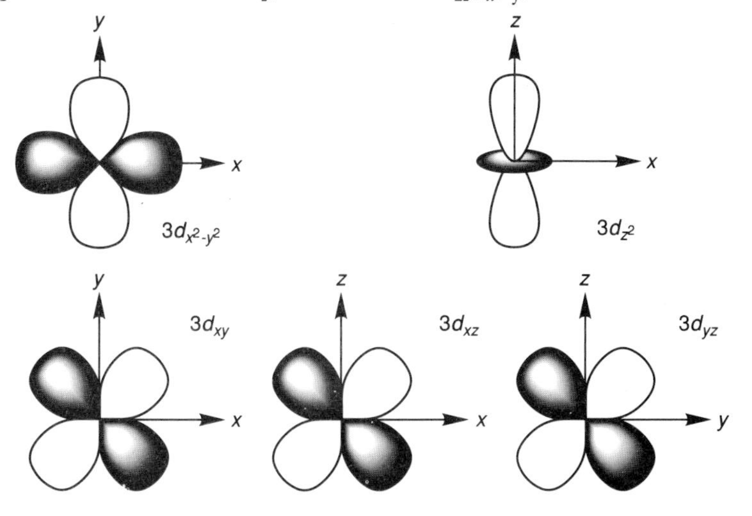

Fig. 1.23 The five *d* orbitals. Note the changes in axis labelling for each orbital.

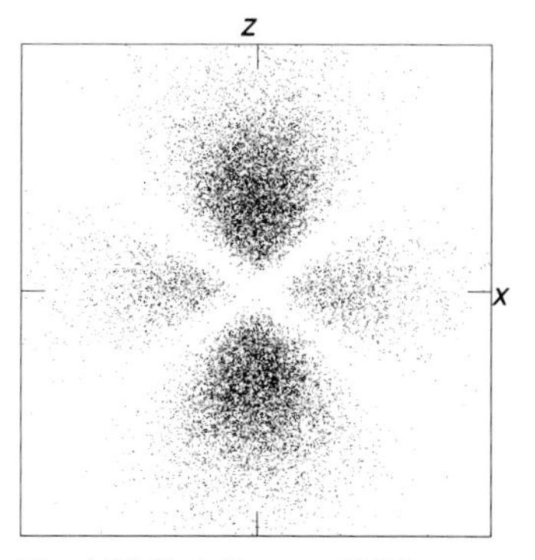

Fig 1.24 Dot diagram (2000 pm square, 15 000 dots) for the hydrogen $3d_{z^2}$ orbital. The apices of two nodal cones intersect at the nucleus, which is at the centre of the square.

1.14 Occupancy of orbitals

Any orbital can accommodate just *two* electrons. It can also hold zero or one electrons, but *never* three or more. While the preceding pages indicate that it is not always appropriate to regard the electron as a little hard ball, certain properties of electrons are best treated as if the electrons spin about an axis passing through a diameter of the electron, rather like the earth spins about an axis every 24 hours. Therefore when there are two electrons in an orbital, one is said to *rotate* or *spin* in one direction and the second in the other. This is unfortunate terminology perhaps but, again, in common usage. The *spin quantum number* differentiates between the two directions of spin. The spin quantum number takes the values $+^1/_2$ or $-^1/_2$. This spin is not to be confused with the notion of electrons rotating about, or orbiting, the nucleus.

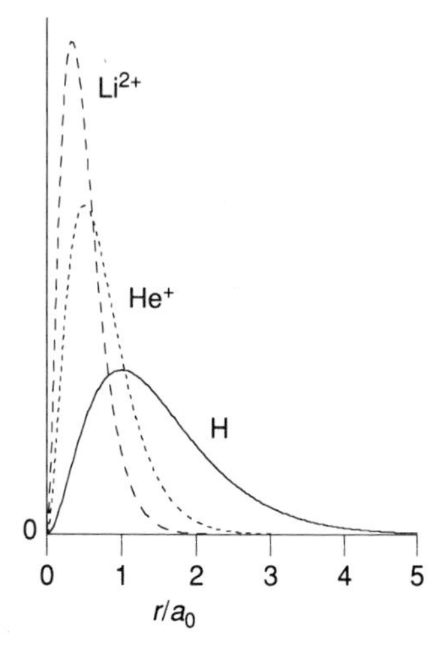

Fig. 1.25 Radial distribution functions for the H, He^+, and Li^{2+} 1*s* orbitals.

1.15 Wave equation for other one-electron atoms

Hydrogen is the only neutral one-electron atom. All other one-electron atoms are positively charged and normally referred to as ions (or more specifically cations). The wave equations for one-electron ions such as He^+ or Li^{2+} are closely related to those of hydrogen. The most important additional factor is the charge on the nucleus. In hydrogen, the nucleus has a charge of +1 from the single proton. In helium, the nucleus has a charge of +2, while in lithium the charge is +3. The effect of the extra positive nuclear charge on the electron is easy to imagine. It will be drawn inwards as a result of the extra attraction. This is expressed in the wave function (Eqn 1.22) for the 1*s* orbital for any one-electron atom which now contains a number Z which is the size of the nuclear charge.

$$\psi_{1s} = \frac{1}{\sqrt{\pi}} Z^{3/2} e^{-Zr} \qquad (1.22)$$

The hydrogen 1*s* wave function is also described by this equation and for hydrogen $Z = 1$. For helium $Z = 2$, for lithium, $Z = 3$, and so on. Figure 1.25 shows plots of ψ^2 and $4\pi r^2\psi^2$ for H, He^+, and Li^{2+} and the contraction arising from the increased nuclear charge is very apparent.

1.16 Building-up principle for many electron atoms

The Schrödinger equation can be solved for atoms other than hydrogen, but not exactly for any system containing more than one electron. This is due to complications arising from electron–electron repulsions. Although the hydrogen wave function solutions are technically only suitable for one-electron atoms, fortunately the hydrogen orbital wave functions are adaptable for many-electron atoms. The justification for adapting the one-electron functions for many-electron systems is that the results are good.

The electronic configuration of an element is a listing of the occupancy of the various atomic orbitals. The process by which the electronic configuration is derived for ever-increasing numbers of electrons is the *building-up principle* also called the *Aufbau principle*.

Ground state hydrogen has the electron in an orbital for which $n = 1$, $l = 0$, and $m_l = 0$. That is, the electron is an a 1*s* orbital. This is denoted $1s^1$. The reason the single electron normally resides in the 1*s* orbital is that the energy of an electron within the 1*s* orbital is at the lowest possible energy. Electrons in any other orbital are at a higher energy.

What about helium? The 1*s* orbital can hold two electrons. The 1*s* orbital is still the lowest energy orbital, therefore the second electron also resides in the 1*s* orbital. It is more favourable to place the second electron in the 1*s* orbital and take an electron–electron repulsion penalty than to place the second electron in some higher energy orbital such as the 2*s* orbital. Note that the energy of an electron in the helium 1*s* orbitals is not the same as the energy of an electron in a hydrogen 1*s* orbital because of the greater nuclear charge.

The values of n, l, and m_l are the same for the second helium electron, but the spin quantum number differs for the two electrons. One electron takes the value $+^1/_2$ while the second takes the value $-^1/_2$. The spins are opposite in direction. The electronic configuration of helium is written as $1s^2$. This is represented on an energy level diagram (Fig. 1.26) using half-arrows. An electron with one spin quantum number is said to be 'spin-up' while an electron with the other spin quantum number is said to be 'spin-down'.

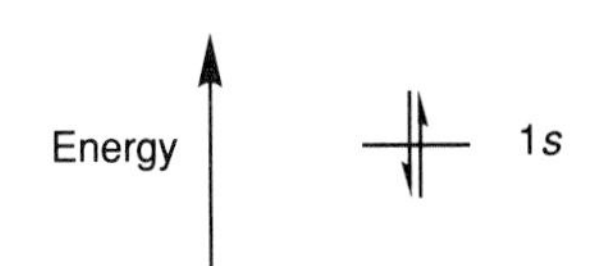

Fig. 1.26 An energy level plot for helium.

The electronic configuration of helium conveniently illustrates the *Pauli exclusion principle* which simply states that no two electrons in any one atom are allowed the same set of four quantum numbers. Each electron in any given atom *must* have a unique address. In the case of helium for instance, the two $1s$ electrons cannot both be spin up (Fig. 1.27), since this would require the spin quantum numbers to be the same.

Fig. 1.27 This configuration for He is forbidden by the Pauli exclusion principle.

Now consider the third element, lithium (Fig 1.28). Since the first shell is fully occupied by two electrons, the third *must* go into an orbital for which $n = 2$ (since the $n = 3$ orbitals are all higher in energy than all the $n = 2$ orbitals). At first sight it could go into either the $2s$ or one of the $2p$ orbitals since for a hydrogen-like atom we have already seen that these orbitals are degenerate. However for a many-electron atom the orbitals are *not* degenerate and the $2s$ orbital is lower in energy than the $2p$. This means the $2s$ orbital is occupied first. The reason for this is a consequence of *orbital screening*, or shielding. The electronic configuration of lithium is $1s^22s^1$.

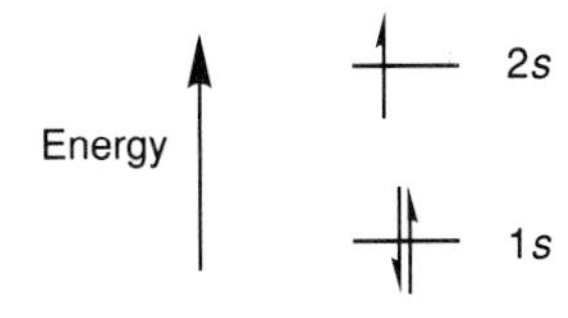

Fig. 1.28 An energy level diagram for ground state lithium ($1s^22s^1$).

1.17 Orbital shielding

Consider the third electron in a lithium atom. It is affected by the charge on the nucleus (+3). However it is also repelled by the charge on the inner two electrons. Figure 1.29 shows the radial distribution functions for the lithium $1s$, $2s$, and $2p$ orbitals. It is clear that the electrons in the $1s$ orbital are much closer to the nucleus than any electrons in the $2s$ or $2p$ orbital. The two inner electrons in the $1s$ orbital *screen* the third electron from the full effect of the nuclear +3 charge. The difference between the full nuclear charge, Z, and the screening effect of the inner two electrons is called the *effective nuclear charge*, or Z_{eff}. In general, for any many-electron atom, any particular electron will always be screened from the nucleus to some extent by the remaining electrons.

$$Z_{eff} = Z - \text{screening constant} \quad (1.22)$$

Figure 1.29 shows the radial density functions of the $2s$ and $2p$ orbitals superimposed on the same plot for the $1s$ orbital in the lithium atom. An electron in the $2s$ orbital is screened from the nucleus by a different amount than an electron in a $2p$ orbital. Examine the region closer to the nucleus, where r is small. There is a significant amount of electron density very close to the nucleus for the $2s$ orbital which is *effectively inside* the $1s$ orbital. In other words the $2s$ orbital intrudes *inside* the volume of space occupied by the $1s$ orbital. There is no such effect for the $2p$ orbital, although there is some penetration. The upshot this is that the $1s$ electrons shield an electron in the $2s$ orbital *less* effectively than they shield an electron in a $2p$ orbital. The effective nuclear charge, Z_{eff} experienced by an electron in a $2s$ orbital is *greater* than that experienced by an electron in a $2p$ orbital. The energy of an

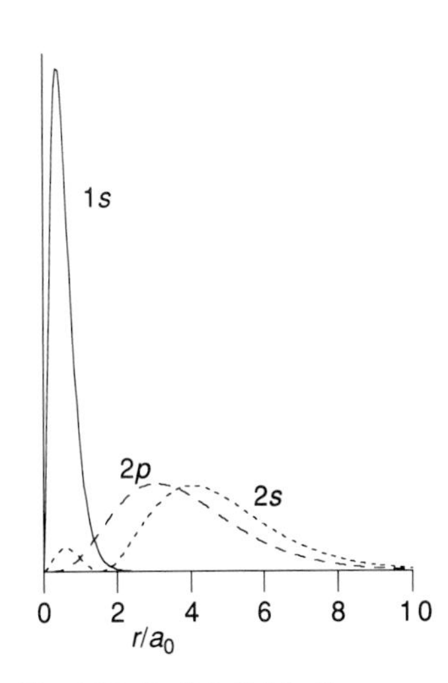

Fig 1.29 Radial distribution functions for lithium $1s$, $2s$, and a $2p$ orbital.

electron in an orbital is dependent upon nuclear charge and calculations show that the energy of the third lithium electron is lower if it is in the $2s$ orbital rather than in a $2p$ orbital. The preferred configuration is therefore $1s^22s^1$ rather than $1s^22p^1$.

The next element is beryllium. Again it is energetically more favourable to place the last electron spin-paired in the $2s$ orbital rather than in one of the $2p$ orbitals. The configuration of beryllium is $1s^22s^2$.

1.18 Hund's rule of maximum multiplicity

At boron, the fifth element, the $2s$ orbital is completely filled. Therefore the next electron goes into one of the $2p$ orbitals. It does not matter which, as each is degenerate. The configuration of boron is therefore $1s^22s^22p^1$. But what happens for carbon, the sixth element: $1s^22s^22p^2$. How are the two electrons in the $2p$ orbitals arranged?

The three configurations in Fig 1.30 are all allowed under the Pauli exclusion rule, in that their quantum number addresses are unique. The dilemma is resolved by *Hund's rule of maximum multiplicity*. This states that the lowest energy configuration for a partially occupied set of degenerate orbitals is that in which as many electrons as possible reside in different orbitals with all their spins in the same direction ('spin parallel'). This is configuration *a* in Fig 1.30. In configuration *b*, extra energy is required to pair the two electrons in one orbital. In configuration *c*, the spins are not parallel. Hund's rule can be rationalized with detailed mathematics, it is more than an empirical observation.

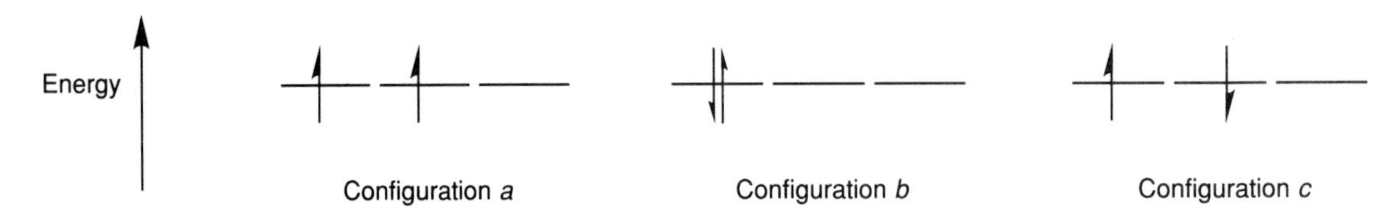

Fig. 1.30 Three possible configurations for two electrons occupying a set of three degenerate $2p$ orbitals.

1.19 Calculation of the effective nuclear charge

The energy of an electron in an orbital and the orbital size are dependent upon the effective nuclear charge, Z_{eff}. An effective discussion on bonding requires a knowledge of the ways in which atomic orbitals interact, and clearly the size and energies of electrons in atomic orbitals will be very important. The usual way to derive the effective nuclear charge is to follow a set of empirical rules designed to put a number to the screening effect of other electrons in the atom. The simplest set of rules are *Slater's rules*, a numerical recipe to derive the effective screening of electrons in an orbital by other electrons. The procedure is to first write the electronic configuration using the following groups of orbitals, and then to apply the following rules:

$$[1s][2s2p][3s3p][3d][4s4p][4d][4f][5s5p][5d][5f]\ldots$$

1. If an electron is in a $[ns,np]$ group, then all the electrons in a group to the *right* contribute zero to the shielding.
2. If an electron is in a $[ns,np]$ group, then any other electrons in that same group of orbitals contribute 0.35 to the shielding. Electrons in the $1s$ orbital are an exception and contribute 0.3.
3. Every electron in a group immediately to the *left* contributes 0.85.
4. Every electron in a group two groups or more to the *left* contribute 1.0.
5. When considering electrons in a $[nd]$ or $[nf]$ group, rules 1 and 2 operate but all electrons in groups to the left of the $[nd]$ or $[nf]$ group contribute 1.0.

As with any set of rules, the easiest way to understand them is to examine some examples. Consider the calculation for one of the $2p$ electrons in carbon. Carbon is written as $[1s]^2[2s2p]^4$ for the purposes of the calculation.

- Rule 1 applies but there are no groups to the right so there is no contribution.
- Rule 2 applies. The other three electrons contribute $3 \times 0.35 = 1.05$.
- Rule 3 applies. The two $1s$ electrons contribute $2 \times 0.85 = 1.7$.
- Rule 4 does not apply. There are no groups two to the left of the $[2s2p]$ group.
- Rule 5 does not apply.

The total contribution to screening is therefore $1.05 + 1.7 = 2.75$. Since $Z = 6$ for carbon, then $Z_{eff} = 6 - 2.75 = 3.25$. The other entries in the table of Slater effective nuclear charges are built up in a similar fashion (Table 1.6). The numbers used in these calculations are set to the values indicated because they work (that is, they give decent answers when calculating energies), not for any particular profound philosophical or mathematical reason. However, they are too approximate to predict, for instance, that the electron in a lithium $2s$ orbital is at a lower energy than if it were in a $2p$ orbital. Some of the rules are easy to rationalize. For instance it should be reasonably clear that any electrons in, say, the $1s$ orbital will totally screen electrons in, say a $3p$ orbital and the contribution to screening upon a $3p$ electron from a $1s$ electron is therefore 1.0. In cases where an outer orbital intrudes into an inner orbital (Fig 1.29), then the screening of electrons in the outer orbital by those in the inner is not complete and so the contribution to screening is something less than 1.0.

Even allowing for the fact that the Slater rules are very approximate, one thing is immediately obvious. Consider the numbers for the $2p$ orbitals. It is clear that the effective nuclear charge experienced by the $2p$ electrons increases sharply on going from left to right in the periodic table. As the effective nuclear charge *increases*, the orbitals *contract*, which is why fluorine atoms are smaller than boron atoms. A more elaborate set of rules (Clementi–Raimondi rules, Table 1.7) gives better values of Z_{eff}, but again they are just that, rules.

Table 1.6. Slater effective nuclear charges, Z_{eff}

Z	$1s$	$2s$	$2p$
1 (H)	1.0		
2 (He)	1.7		
3 (Li)	2.7	1.3	
4 (Be)	3.7	1.95	
5 (B)	4.7	2.6	2.6
6 (C)	5.7	3.25	3.25
7 (N)	6.7	3.9	3.9
8 (O)	7.7	4.55	4.55
9 (F)	8.7	5.2	5.2
10 (Ne)	9.7	5.85	5.85

Table 1.7. Clementi–Raimondi effective nuclear charges, Z_{eff}

Z	$1s$	$2s$	$2p$
1 (H)	1.0		
2 (He)	1.69		
3 (Li)	2.69	1.28	
4 (Be)	3.68	1.91	
5 (B)	4.68	2.58	2.42
6 (C)	5.67	3.22	3.14
7 (N)	6.66	3.85	3.83
8 (O)	7.66	4.49	4.45
9 (F)	8.65	5.13	5.10
10 (Ne)	9.64	5.76	5.76

1.20 Electronic structure and the periodic table

The four quantum numbers n, l, m_l, and m_s are sufficient to describe the electronic configurations of all the elements. The electronic configurations of all the elements are rationalized using the above rules. The electronic configurations of the first 36 elements are listed in Table 1.8.

These are *gas phase* configurations of *neutral atoms*, electronic configurations in other environments such as transition metal compounds need not be the same. There are some unexpected results. Starting at

Table 1.8. Electronic structures of elements 1–36 (gas phase)

Z	Symbol	Electronic configuration	Shorthand form
1	H	$1s^1$	$1s^1$
2	He	$1s^2$	$1s^2$
3	Li	$1s^2 2s^1$	[He]$2s^1$
4	Be	$1s^2 2s^2$	[He]$2s^2$
5	B	$1s^2 2s^2 2p^1$	[He]$2s^2 2p^1$
6	C	$1s^2 2s^2 2p^2$	[He]$2s^2 2p^2$
7	N	$1s^2 2s^2 2p^3$	[He]$2s^2 2p^3$
8	O	$1s^2 2s^2 2p^4$	[He]$2s^2 2p^4$
9	F	$1s^2 2s^2 2p^5$	[He]$2s^2 2p^5$
10	Ne	$1s^2 2s^2 2p^6$	[He]$2s^2 2p^6$
11	Na	$1s^2 2s^2 2p^6 3s^1$	[Ne]$3s^1$
12	Mg	$1s^2 2s^2 2p^6 3s^2$	[Ne]$3s^2$
13	Al	$1s^2 2s^2 2p^6 3s^2 3d^1$	[Ne]$3s^2 3p^1$
14	Si	$1s^2 2s^2 2p^6 3s^2 3p^2$	[Ne]$3s^2 3p^2$
15	P	$1s^2 2s^2 2p^6 3s^2 3p^3$	[Ne]$3s^2 3p^3$
16	S	$1s^2 2s^2 2p^6 3s^2 3p^4$	[Ne]$3s^2 3p^4$
17	Cl	$1s^2 2s^2 2p^6 3s^2 3p^5$	[Ne]$3s^2 3p^5$
18	Ar	$1s^2 2s^2 2p^6 3s^2 3p^6$	[Ne]$3s^2 3p^6$
19	K	$1s^2 2s^2 2p^6 3s^2 3p^6 4s^1$	[Ar]$4s^1$
20	Ca	$1s^2 2s^2 2p^6 3s^2 3p^6 4s^2$	[Ar]$4s^2$
21	Sc	$1s^2 2s^2 2p^6 3s^2 3p^6 4s^2 3d^1$	[Ar]$4s^2 3d^1$
22	Ti	$1s^2 2s^2 2p^6 3s^2 3p^6 4s^2 3d^2$	[Ar]$4s^2 3d^2$
23	V	$1s^2 2s^2 2p^6 3s^2 3p^6 4s^2 3d^3$	[Ar]$4s^2 3d^3$
24	Cr	$1s^2 2s^2 2p^6 3s^2 3p^6 4s^1 3d^5$	[Ar]$4s^1 3d^5$
25	Mn	$1s^2 2s^2 2p^6 3s^2 3p^6 4s^2 3d^5$	[Ar]$4s^2 3d^5$
26	Fe	$1s^2 2s^2 2p^6 3s^2 3p^6 4s^2 3d^6$	[Ar]$4s^2 3d^6$
27	Co	$1s^2 2s^2 2p^6 3s^2 3p^6 4s^2 3d^7$	[Ar]$4s^2 3d^7$
28	Ni	$1s^2 2s^2 2p^6 3s^2 3p^6 4s^2 3d^8$	[Ar]$4s^2 3d^8$
29	Cu	$1s^2 2s^2 2p^6 3s^2 3p^6 4s^1 3d^{10}$	[Ar]$4s^1 3d^{10}$
30	Zn	$1s^2 2s^2 2p^6 3s^2 3p^6 4s^2 3d^{10}$	[Ar]$4s^2 3d^{10}$
31	Ga	$1s^2 2s^2 2p^6 3s^2 3p^6 4s^2 3d^{10} 4p^1$	[Ar]$4s^2 3d^{10} 4p^1$
32	Ge	$1s^2 2s^2 2p^6 3s^2 3p^6 4s^2 3d^{10} 4p^2$	[Ar]$4s^2 3d^{10} 4p^2$
33	As	$1s^2 2s^2 2p^6 3s^2 3p^6 4s^2 3d^{10} 4p^3$	[Ar]$4s^2 3d^{10} 4p^3$
34	Se	$1s^2 2s^2 2p^6 3s^2 3p^6 4s^2 3d^{10} 4p^4$	[Ar]$4s^2 3d^{10} 4p^4$
35	Br	$1s^2 2s^2 2p^6 3s^2 3p^6 4s^1 3d^{10} 4p^5$	[Ar]$4s^2 3d^{10} 4p^5$
36	Kr	$1s^2 2s^2 2p^6 3s^2 3p^6 4s^2 3d^{10} 4p^6$	[Ar]$4s^2 3d^{10} 4p^6$

potassium (Z = 19), the 4*s* orbital fills before the 3*d* level. The energy of an electron in the potassium 4*s* orbital is lower than one in a 3*d* orbital. The situtation for scandium (Z = 21) is more complex. The ground state configuration of the neutral atom is Ar]$4s^2 3d^1$. Experimentally, however, the 4*s* level is *above* the 3*d* level and yet it fills *before* the 3*d* level, apparently in contravention of the building-up principle. There is no simple explanation of this phenomenon, which is to do with electron–electron repulsions, and the interested reader should consult a more advanced text. Note also that chromium is [Ar]$4s^1 3d^5$ rather than [Ar]$4s^2 3d^4$ and that copper is [Ar]$4s^1 3d^{10}$ rather than [Ar]$4s^2 3d^9$. These configurations are adopted rather than those expected because the total energy of the system in each case is less.

The electronic structures of Table 1.6 and those of the heavier elements are the basis of the periodic table (inside back cover). The periodic table is based upon the realization that reactivity is periodic. Elements whose chemistry is related appear in columns. Their chemistries are related because their outermost orbital electronic configurations are similar. Thus, the outermost orbital electronic configurations of all the halogens (F, Cl, Br, and I) is $s^2 p^5$.

The standard periodic table is not a unique arrangement of the elements. Other arrangements are possible and have their own advantages and disadvantages over the standard form. A good mnemonic aid to remembering the order in which the orbitals are filled is shown in Fig. 1.31 and is read from left to right and top to bottom. This tracks a modernized form of the *Janet periodic table*, a table preferred by some to the standard form.

1.21 Lewis structures of atoms

It is often inconvenient to write down the complete electronic structure for an atom when considering bonding, largely because it is only the valence electrons that are involved in reactions. For sulphur, for instance, it is convenient to write [Ne]$3s^2 3p^4$ as the electronic structure, or even just $s^2 p^4$, rather than $1s^2 2s^2 2p^6 3s^2 3p^4$. There are six valence electrons, that is, there are

			1*s*
			2*s*
		2*p*	3*s*
		3*p*	4*s*
	3*d*	4*p*	5*s*
	4*d*	5*p*	6*s*
4*f*	5*d*	6*p*	7*s*
5*f*	6*d*	7*p*	8*s*

																														1 H	2 He
																														3 Li	4 Be
																								5 B	6 C	7 N	8 O	9 F	10 Ne	11 Na	12 Mg
																								13 Al	14 Si	15 P	16 S	17 Cl	18 Ar	19 K	20 Ca
														21 Sc	22 Ti	23 V	24 Cr	25 Mn	26 Fe	27 Co	28 Ni	29 Cu	30 Zn	31 Ga	32 Ge	33 As	34 Se	35 Br	36 Kr	37 Rb	38 Sr
														39 Y	40 Zr	41 Nb	42 Mo	43 Tc	44 Ru	45 Rh	46 Pd	47 Ag	48 Cd	49 In	50 Sn	51 Sb	52 Te	53 I	54 Xe	55 Cs	56 Ba
57 La	58 Ce	59 Pr	60 Nd	61 Pm	62 Sm	63 Eu	64 Gd	65 Tb	66 Dy	67 Ho	68 Er	69 Tm	70 Yb	71 Lu	72 Hf	73 Ta	74 W	75 Re	76 Os	77 Ir	78 Pt	79 Au	80 Hg	81 Tl	82 Pb	83 Bi	84 Po	85 At	86 Rn	87 Fr	88 Ra
89 Ac	90 Th	91 Pa	92 U	93 Np	94 Pu	95 Am	96 Cm	97 Bk	98 Cf	99 Es	100 Fm	101 Md	102 No	103 Lr	104 Rf	105 Ha															

Fig. 1.31 A chart to remember the order in which atomic orbitals are filled for gas phase neutral atoms and a modernized form of the Janet periodic table.

Core electrons: innermost electrons corresponding to the electron count of the previous noble gas.

Valence electrons: all the electrons that are not core electrons.

Gilbert N. Lewis: 1875–1947

six electrons outside the neon core, neon being the previous noble gas.

The valence electron structure of sulphur can be represented by a Lewis structure ·S̤̈· in which the six dots represent the six valence electrons. The next available closed shell structure atom after sulphur is argon. Argon has two more electrons than sulphur meaning that S^{2-} (:S̤̈:) has the same electronic configuration as Ar. The Lewis atom structures of the first few elements (Fig. 1.32) make it is clear why their structures are periodic.

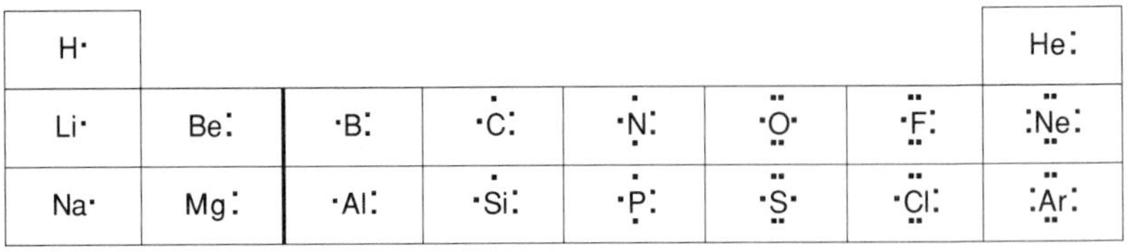

H·							He:
Li·	Be:	·B:	·Ċ:	·Ṅ:	·Ö:	·F̤̈:	:N̤̈e:
Na·	Mg:	·Al:	·Ṡi:	·Ṗ:	·S̤̈·	·C̤̈l:	:Ä̤r:

Fig 1.32 Lewis structures of the first eighteen elements.

1.22 Exercises and problems

1. Calculate the most likely radius at which an electron will be found in a H, He^+ , and Li^{2+} 3*s* orbitals.
2. Calculate the radii at which radial nodes appear for H, He^+, and Li^{2+} 3*s* orbitals.
3. Calculate how much energy is required to remove the electron from a hydrogen atom in which the electron occupies the $n = 3$ level.
4. Calculate the frequencies of the first five lines of the Lyman, Balmer, Paschen, Brackett, and Pfund atomic emission lines of hydrogen. Plot these on a scale to simulate the emission line spectrum of hydrogen.
5. Making use of appropriate books, determine which frequency ranges correspond to each of the visible colours. Which of the emission lines calculated above fall into the visible part of the spectrum?
6. The wave equation for a hydrogen $2p_z$ orbital with r expressed in atomic units is given by:

$$\psi_{2p_z} = \frac{1}{4\sqrt{2\pi}}(\cos\theta)re^{-r/2}$$

The line connecting the point in question to the nucleus makes an angle θ with the z axis. Plot graphs of this function and $\psi_{2p}{}^2$ along the z axis ($\theta = 0$) out to ± 10 atomic units. What happens to these functions at $\theta = 90°$?

7. Extend Table 1.6 (Slater effective nuclear charges) up to $Z = 36$.
8. Determine the number of orbitals in each subshell and shell for $n = 1$ to 6. Determine a simple formula to relate the number of orbitals in a shell to n.
9. This book shows the standard periodic table (inside back cover) and the *Janet periodic table* (Fig. 1.31). Find an example of another periodic table classification, write it out, and determine the relationships between the structures of the three tables.

2 Lewis molecular structures

2.1 Covalent and ionic bonding

There are two principal types of non-metallic chemical bonding, *covalent* and *ionic*. Ionic bonding occurs when a compound such as common salt, NaCl, adopts a lattice structure consisting of a regular array of positively charged ions (cations) and negatively charged ions (anions). The ions are held together largely by electrostatic forces between the cations and anions.

In 1916, Lewis suggested that a *covalent* bond is where two atoms are held together by a pair of electrons located between the two atoms. The two electrons are *shared* by the two atoms. For electronic bookkeeping purposes, in most molecules, each of the two atoms provides one electron each to the bond. This is the basis of the simplistic but useful '*dot and cross*' bonding representation. Since this concept was developed by Lewis, such structures are often called Lewis structures. The essence of this treatment is that atoms involved in covalent bonding should attain the electronic configuration of the nearest noble gas. This is also true for ionic structures. In NaCl one electron is transferred from sodium to chlorine to make Na^+ and Cl^- ions. Each ion attains the electronic structure of the nearest noble gas and it is these ions which make up the salt lattice structure.

Noble, or *rare*, *gases*: He, Ne, Ar, Kr, Xe, Rn

2.2 Lewis bonding of H_2

The Lewis structure of H_2 is very simple. Each hydrogen atom has one electron. In the molecule H_2, each hydrogen attains the electronic configuration of the next noble gas, in this case the helium duplet, by sharing the two electrons. This is represented as H:H, or more simply H—H. This representation tells us that there are two electrons located between the two hydrogen nuclei, and that these two electrons form the bond holding the nuclei together. The interactions in H_2 are represented in Fig 2.1. Don't forget that the electrons are not really point charges. There are a number of interactions. The force interaction between two point charges varies with the square of the distance between them. The two protons are positively charged and repel each other. The two electrons are negatively charged and repel each other. However this is more than offset by the four proton–electron attractions. These six interactions represent the net bonding in the system.

It is immediately apparent that as the number of atoms in a molecule increases and as the number of electrons increases, the situation become really rather complex. This model is flawed. Neither the protons or electrons are point charges. Nevertheless, the model is instructive.

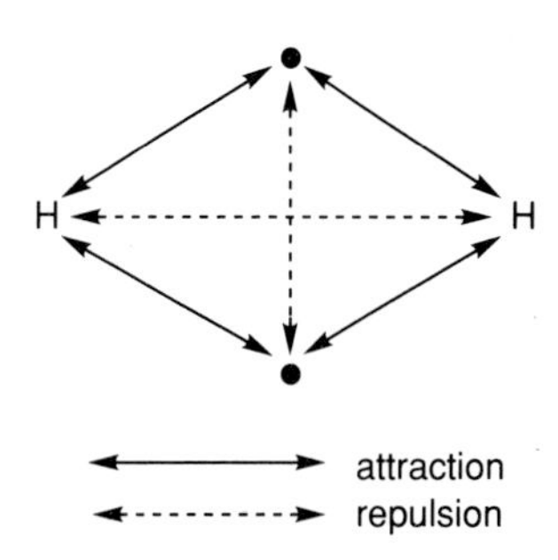

Fig 2.1 The interactions between the two hydrogen nuclei and the two electrons of H_2.

2.3 The octet rule and diatomic molecules

Lewis realized that most of the second period elements, in particular, are surrounded in covalent compounds by eight electrons. They do this by sharing electrons with neighbouring atoms. The number eight is the magic number because all the valence orbitals are fully occupied, although at the time the octet rule was first recognized, the reasons were not appreciated.

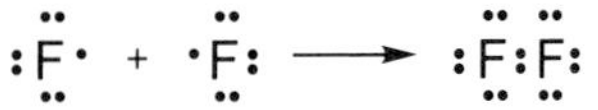

:F· + ·F: ⟶ :F:F:

The fluorine atom has seven valence electrons and only requires one extra electron to attain the neon electronic structure. One way it can do this is by sharing an electron with another fluorine atom. The resulting molecule, written F—F, demonstrates another point. It has six pairs of electrons that are *not* shared with other atoms. Such pairs of electrons are *lone pairs*. Not unreasonably, the lone pairs of electrons on adjacent fluorine atoms repel each other. The resulting interactions partially negate the bonding energy from the shared pair of electrons and the result is that the two fluorine atoms in F—F are very weakly bound. This is one of the reasons that F_2 is so reactive.

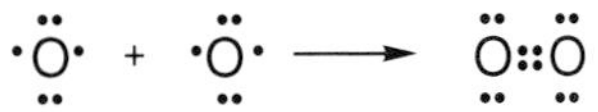

Lone pair: a pair of valence electrons not shared with another atom

·O· + ·O· ⟶ O::O

Oxygen with its six valence electrons requires *two* electrons to attain the octet structure of neon. One way the oxygen atom can do this is to share *two* electrons with a neighbouring oxygen atom. This places *two* shared pairs of electrons between the two oxygen nuclei. This introduces the concept of *bond order*. Two electrons located between two atoms are defined as forming a *single bond*. Two pairs of electrons between two atoms constitute two bonds, that is, a *double bond*. The bond order in dioxygen is 2 and dioxygen is written O=O. Each oxygen atom has two lone pairs.

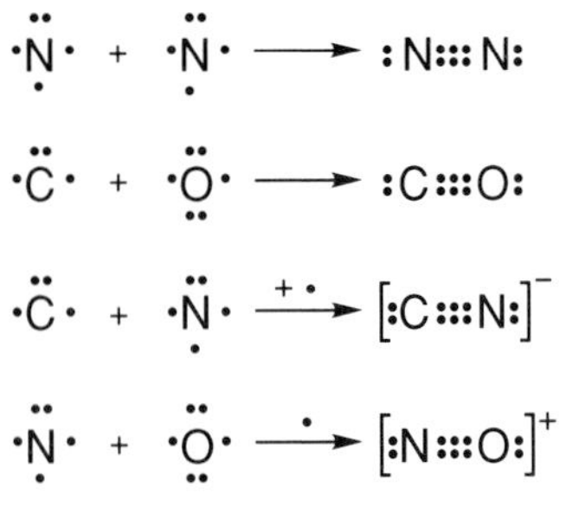

·N· + ·N· ⟶ :N:::N:

·C· + ·O· ⟶ :C:::O:

·C· + ·N· ⟶ (+ ·) [:C:::N:]$^-$

·N· + ·O· ⟶ (− ·) [:N:::O:]$^+$

Nitrogen has five valence electrons and satisfaction of the octet rule therefore requires that it share *three* electrons with a neighbouring nitrogen atom. This means that dinitrogen is written N≡N to show the *triple bond* and the bond order is 3. Each nitrogen atom has just one lone pair.

The impression should now be forming that constructing Lewis structure diagrams is a matter of determining the total number of electrons possessed by the two atoms and arranging them so that the octet rule (or the duplet rule if one of the atoms is hydrogen) is satisfied for each atom.

Thus, carbon monoxide contains carbon with 4 valence electrons and oxygen with 6, total 10. This is the same number of electrons possessed by the N≡N molecule (structure :N≡N:) and carbon monoxide has its ten electrons arranged in the same way (structure :C≡O:). Charged species are treated simply by adding a dot in the case of a monoanion, or subtracting a dot in the case of a monocation. Therefore cyanide, $C{\equiv}N^-$, and the nitrosonium cation, $N{\equiv}O^+$ are both 10–electron systems written with a triple bond between the two atoms. Molecules with the same number of component atoms and the same total electron count, such as N_2, CO, CN^-, and NO^+ are said to be *isoelectronic*. One would expect there to be at least some similarities in their chemistries.

2.4 Polar bonds and ionic bonding

Homonuclear diatomic molecule such as N_2, O_2, and F_2 are symmetrical and therefore there cannot be a charge imbalance between the two atoms. The sharing of electrons in this way is 'pure' covalent bonding. Diatomic molecules such as HF are asymmetric and there is no reason, except by accident, for there *not* to be a charge imbalance between the two atoms. An *electronegativity scale* is a set of numbers which relates the power of an atom to attract electrons towards itself within a molecule. Elements which do this successfully are said to be very electronegative. Elements that do not do this well are weakly electronegative, or electropositive. There are a number of numerical electronegativity scales, such as the *Pauling scale* or the *Mulliken scale*, in common use. Elements towards the right and top of the standard periodic table are the most electronegative. Elements towards the bottom and left are least electronegative, or *electropositive*.

Therefore in a diatomic molecule such as H—F, the fluorine, being more electronegative, attracts more than its fair share of electron density. It does this at the expense of the less electronegative hydrogen. The consequence of this is that there is more electron density at the fluorine end of the molecule than at the hydrogen end. The bond is *polar*.

$\overset{\delta+}{H}—\overset{\delta-}{F} \equiv H—F$ (with arrow ⟶ pointing to F)

Two ways to denote that the H—F bond is polar.

Consider lithium fluoride, LiF. One way to satisfy the noble gas rule for *both* atoms is to completely transfer an electron from the very electropositive lithium to the very electronegative fluorine. Fluorine attains the electronic configuration of neon while lithium attains the configuration of helium. The Lewis representation of such an electron transfer is $[Li]^+[:\ddot{F}:]^-$. Removal of an electron from lithium corresponds to *ionization* and addition of an electron to fluorine is the reverse of *electron affinity*. The energy penalties for electron transfer are more than compensated by the electrostatic attraction between the Li^+ cation and F^- anion. This is *ionic bonding*. Ionic bonding is favoured, in particular, for those diatomic cases where one of the atoms is electropositive and the other very electronegative.

$Li\cdot + :\ddot{F}\cdot \longrightarrow [Li]^+[:\ddot{F}:]^-$

Ionization enthalpy: the enthalpy change in the reaction $M \rightarrow M^+$.
Electron affinity: the enthalpy change in the reaction $M^- \rightarrow M$.

In practice, however, the transfer of electrons is never complete. If a lithium cation is placed close to a fluorine anion, the electron clouds are distorted by the neighbouring charge. The electron cloud around the anion is distorted towards the cation. This amounts to a partial transfer of electron density back to the cation along the internuclear axis. Placing some charge density along the internuclear axis constitutes partial covalent bonding The consequence of this is that it is impossible to have a *pure* ionic bond since some distortion in the anion charge cloud always occurs. Looked at this way, diatomic ionic structures are an extreme form of polar bonding in which the bond is so polarized that electrons are almost completely transferred to the more electronegative atom. In practice, one can place any bond on a 'spectrum' of bond types which varies from 'pure' covalent bonding to 'pure' ionic bonding with pretty well all molecules lying between the extremes.

Pure ionic — very polar — slightly polar — pure covalent

2.5 The octet rule and polyatomic molecules

Extension of the Lewis model to polyatomic structures is relatively straightforward. One potential problem is knowing the connectivity of the molecule. Given an empirical formula C_2H_3N there is no way of knowing whether the structure should be MeC≡N: (methyl cyanide), MeN≡C: (methyl isocyanide), or some other structure. Further, Lewis structures give no clue as to the geometry of the molecule in question.

The procedure for writing down the Lewis structure for a polyatomic molecule is to calculate the total number of valence electrons for all the atoms in a molecule, make any necessary compensation for charge, and arrange all the electrons around the component atoms (arranged in the correct connectivity) so that the octet rule (or duplet rule for hydrogen) is satisfied for each (Fig. 2.2). Note that the Lewis representation of ethene successfully accounts for the double bond. Atoms in other molecules such as NH_3 do not share all their electrons. Nitrogen has a pair of electrons it retains for itself. Such a pair is termed a *lone pair*. Note that nitrogen still has an octet of electrons.

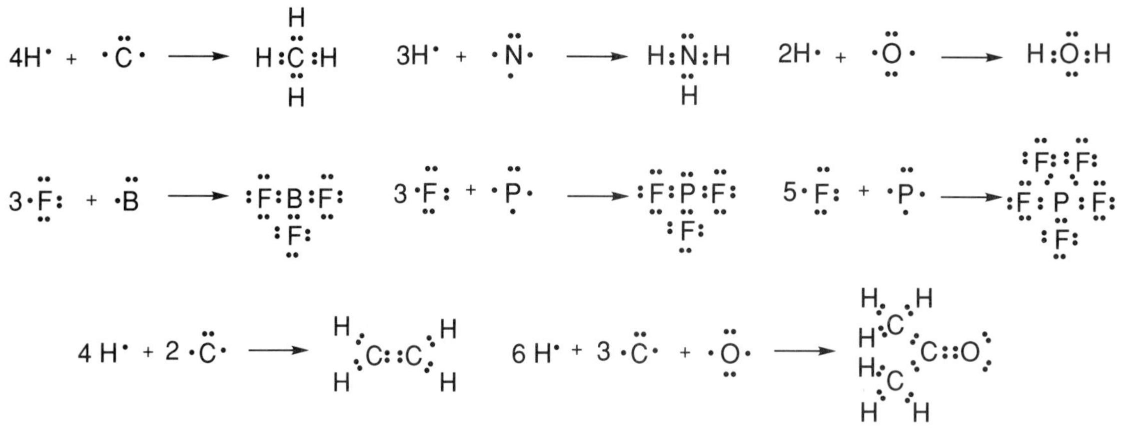

Fig. 2.2 Some examples of Lewis structures.

There are a very few exceptions to the octet rule for the second period elements. In BF_3, all the fluorine atoms attain the octet configuration but boron has only six. Molecules of this type are sometimes termed *electron deficient*. The octet rule is frequently adhered to by main group elements below the second period, but there is more scope for exceptions. So for the third period elements, in addition to the 3*s* and 3*p* orbitals there are a number of cases, such as PF_5, where the electron count can expands beyond eight. This is often attributed to participation in the bonding by *d* orbitals.

It is common to denote bonds by a line such as that in F—F. The line corresponds to two electron dots in the diagrams above and is generally a more convenient representation than the dots. The two dots are retained for lone pairs, but lone pairs are often only denoted on the atom of interest in a molecule (Fig. 2.3). So in PF_3, only the phosphorus lone pair is generally

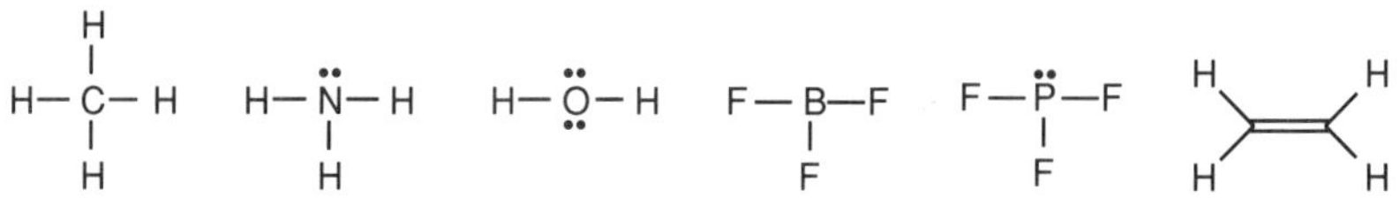

Fig. 2.3 Lewis structures in which only the lone pairs on the central atoms are denoted.

written and the fluorine lone pairs are left from the diagram.

2.6 Dative bonding

Lewis bases: electron donors
Lewis acids: electron acceptors

One way for the boron atom in an electron-deficient molecule such as BF_3 to attain the octet configuration is for it to acquire *two* electrons from another molecule. Ammonia is a good electron pair donor. The nitrogen shares its lone pair with boron in the *adduct* $F_3B{\leftarrow}NH_3$ (Fig. 2.4). The arrow $\rightarrow$ is often, but not always, used to denote that one atom donates both electrons to the bond. A bond in which both electrons originate from one atom is a *dative covalent bond*. In this example, NH_3 is a Lewis base and BF_3 is a Lewis acid. The origin of the term dative is in the Latin *dare*, to give.

In some ways the arrow symbolism is useful in that it might help electron accountancy. In others it is a hindrance in that the arrow suggests that in some way the bond is different from a shared electron-pair bond. It is not, both shared electron bonds and dative electron bonds are simply two-electron bonds. Once a bond *exists*, there is no way to distinguish any characteristic of that bond which is differs between the two types. Note the representation H_3N^+—B^- F_3 is also perfectly legitimate.

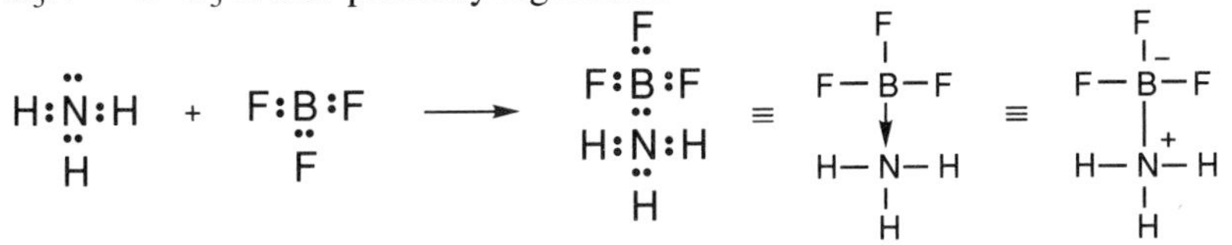

Fig. 2.4 The reaction between the Lewis base NH_3 and the Lewis acid BF_3 to form the adduct $H_3N{\rightarrow}BF_3$.

2.7 Resonance structures

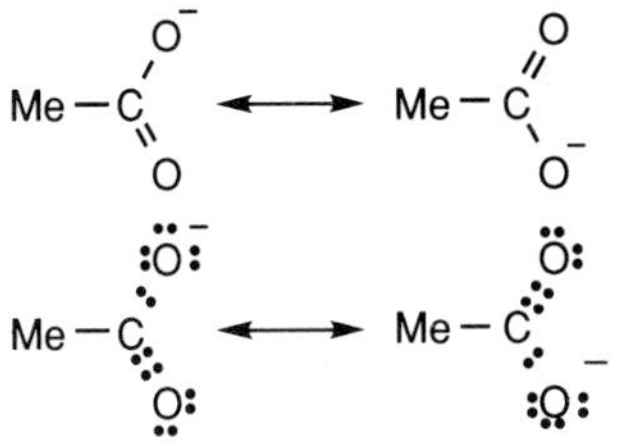

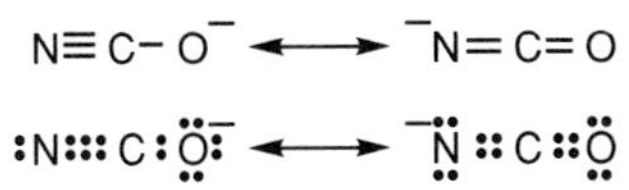

Ethanoic (acetic) acid, $MeCO_2H$, ionizes in solution to some extent to form ethanoate (acetate), $[MeCO_2]^-$. It is possible to write down *two* reasonable dot structures to represent the structure of ethanoate. Which is correct? The answer is neither. Both structures *contribute* to the actual structure. The real structure is said to be a *resonance hybrid* of the two contributor structures. It is not a case of interconversion (so-called 'flickering') between the two structures. The real structure is a single structure, but it is not possible to write down a single Lewis structure to represent it. The use of resonance structures is a common device and it is a good idea to be familiar with them.

For ethanoate the contributor structures are clearly degenerate. In other cases the resonance structures are not degenerate. The cyanate ion, $[CNO]^-$, is one such example and its two most important structures are not degenerate.

2.8 Exercises and problems

1. Write down Lewis structures for H_2O_2, $AlCl_3$, Al_2O_3, CO_2, SO_2, SO_3, HCN, HNC, HNO_3, $HClO_4$, H_2SO_4, SF_4, SF_6, NO_2^-, NO_2, N_2O_4, and NO_2^+.
2. Write down Lewis structures for the most important contributor resonance structures of SO_4^{2-}, O_3, NO_2, NO_2^-, NO_3^-, C_6H_6, and BF_3.

3 Diatomic molecules

3.1 Problems with Lewis representations

A *paramagnetic* molecule is one in which there is one or more unpaired electrons. In *diamagnetic* molecules all the electrons are paired.

For many purposes, Lewis structures are an adequate description of bonding. They do have their problems however. The Lewis structure for O_2 has two oxygen atoms linked by four electrons, two for each bond. All the electrons are apparently paired, either in bonds or as lone pairs. This is an example where the Lewis representation of structures breaks down. It fails to account for the fact that O_2 has two unpaired electrons and is therefore paramagnetic.

The fact that a bonding model, or any other model for that matter, is demonstrably wrong does not mean we must instantly consign it to the rubbish bin. The Lewis dot bonding representation is extremely useful, for instance, in the VSEPR method (Chapter 4) for calculating the shapes of main group compounds.

A more sophisticated model is required to account for the paramagnetism. In outlining such a model it is necessary first to examine the bonding in some simpler diatomic molecules, starting with the simplest possible molecule, H_2^+, the positive ion of the H_2 molecule.

3.2 The simplest diatomic molecule: H_2^+

The Lewis structure of H_2 has two electrons holding the two nuclei together in a single bond. Shining light of an appropriate frequency on the H_2 molecule results in one of the electrons in the H_2 molecule being ejected, with the resultant formation of the transiently stable ion H_2^+. This is the simplest of all molecules. It consists of two protons held together by just a single electron. The bond order in the Lewis structure of H_2^+ is therefore $^1/_2$. Clearly the single electron cannot hold the two protons together as well as the two electrons in H_2. Nevertheless, there is a bond, although it is weaker (255 kJ mol^{-1}) than that in H_2 (430 kJ mol^{-1}). It is also longer with an equilibrium bond length of 106 pm compared to that of H_2 which is 74 pm.

Conceptually the hydrogen atom is made up from a proton and an electron. The H_2^+ molecule is made up from two protons and an electron. The electron in the H atom is located in an orbital centred on the proton. The way forward for H_2^+ is to place the electron in an orbital centred on the two protons. Instead of being in an *atomic* 1*s* orbital, the electron in H_2^+ is in a *molecular orbital*. Unlike the hydrogen atomic orbital which is spherical, the orbital corresponding to the lowest energy level of H_2^+ is cylindrically symmetrical about the H—H axis.

Each point within a molecular orbital has an associated electron density just as atomic orbitals do. Some regions of space contain greater amounts of electron than others. The electron dot density cross-section plot (Fig. 3.1) shows that the bulk, but not all, of the electron density resides in the region between the two nuclei. As for atomic orbitals, this is often inconvenient to represent on a diagram and so again a convention is needed to represent this. The same boundary convention already described for atoms is appropriate. The orbital in H_2^+ that holds the electron is vaguely blimp- or cigar-shaped and generally represented with a boundary sketch (Fig. 3.2).

The plus sign on the orbital shows that the sign of the wave function ψ enclosed by the boundary is positive. This is the same sign convention used for atoms.

So, in molecules, as in atoms, electrons reside in orbitals. An orbital, whether in an atom or a molecule is still an orbital, but their shapes are different and they are associated with more than one atomic nuclei. Orbitals associated with atoms are called *atomic orbitals* and those associated with molecules are called *molecular orbitals*. The description of molecules based on molecular orbitals is called molecular orbital (MO) theory. The resulting molecular orbitals are in many cases spread (*delocalized*) over the entire molecule.

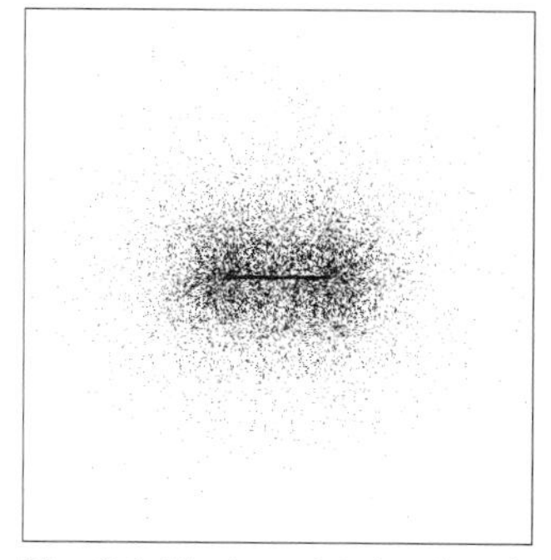

Fig. 3.1 Electron dot density plot (10 000 dots) for the lowest energy molecular orbital in H_2^+. The bond length is set to 106 pm and the plot is a 500 pm square.

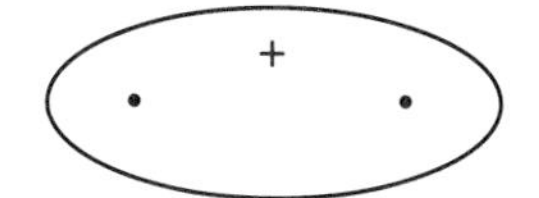

Fig. 3.2 A boundary representation of the molecular orbital on H_2^+ containing the electron.

3.3 Interference in wave phenomena

Sound or light waves can interact in a constructive or destructive fashion. If a pair of loudspeakers is connected to the amplifier so that the terminals on one are reversed relative to the second, the result is a dead sound because as one speaker cone moves out, the other moves in. Some concert halls, and lecture theatres, are acoustically poor. One of the reasons for this is that as the sound bounces around, the reflected waves interact *destructively* with each other. Related effects can be produced with water. Throwing a pebble into a still pond generates a characteristic pattern of ripples. Throwing two pebbles in results in two ripple patterns which overlap with each other, in an idealized example perhaps as in Fig. 3.3 (top).

At any particular point in the pond, two water ripples in phase with each

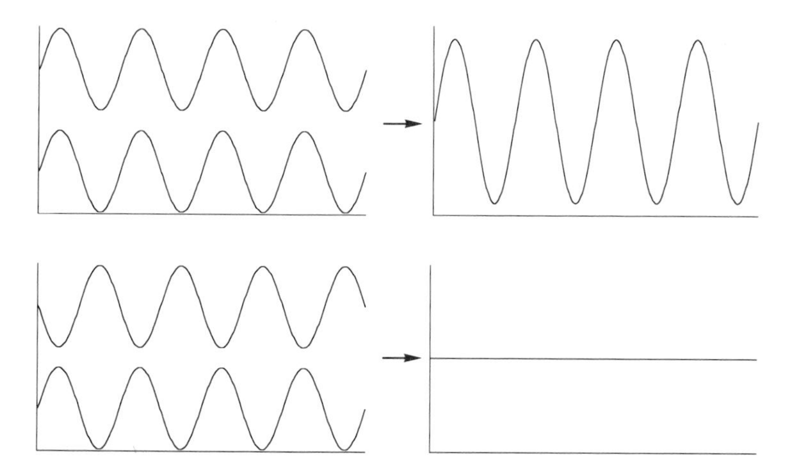

Fig. 3.3 Reinforcement and destruction of waves through interference.

other mix or combine to make a bigger ripple, while if they happen to be out of phase (out of step) with each other, they cancel each other out. The calculation of the pattern representing the reinforcement of two ripples is relatively straightforward. Consider the two waves at the top of Fig 3.3. These waves are identical in terms of their amplitudes and wavelength. The amplitude of the resultant wave at any particular position is simply calculated by *adding* the amplitude of the two component waves. The resultant wave is shown to the right. This operation could be written as in Eqn 3.1.

$$wave + wave \rightarrow 2 \times wave \tag{3.1}$$

$$wave + (-1) \times wave \rightarrow 0 \tag{3.2}$$

If the two waves happen to be out of phase, as in Fig. 3.3 (bottom), the result of this addition is complete destruction and the result is the zero amplitude line to the right. The second (out-of-phase) wave is related to the first in the sense that at every point, the magnitude of the displacement is exactly the same, but that the direction or sign of the wave is opposite. Mathematically this is achieved simply by multiplying the first by –1. This out-of-phase combination could be written as in Eqn 3.2. Effectively this is the same as writing $wave - wave \rightarrow 0$.

3.4 In-phase interaction of 1*s* orbital wave functions

Mathematically the behaviour of electrons in orbitals can be treated as a wave phenomenon, and the wave functions for two interacting orbitals on two different atoms interact with each other constructively or destructively. The wave functions used for atomic orbitals are not sine waves but this doesn't affect the principle of the matter.

The simplest approach to analysing these interactions involves taking *linear combinations* of the constituent atomic orbitals. This involves mathematically mixing the valence atomic orbitals of the respective atoms. In Fig. 3.4, two hydrogen atoms are placed at a bonding distance apart. The 1*s* wave functions interact. As for the water ripples above, the reinforcement at any particular point is calculated simply by adding the amplitude of the wave functions of the two hydrogen nuclei at that point together. The result of this addition is shown to the right.

There is an overlap in the middle (between the two nuclei). This is called the in-phase combination of two 1*s* orbitals. Mathematically, constructing the wave function for the in-phase combination of two H 1*s* orbitals involves adding the two functions together at each point in space (Eqn 3.3).

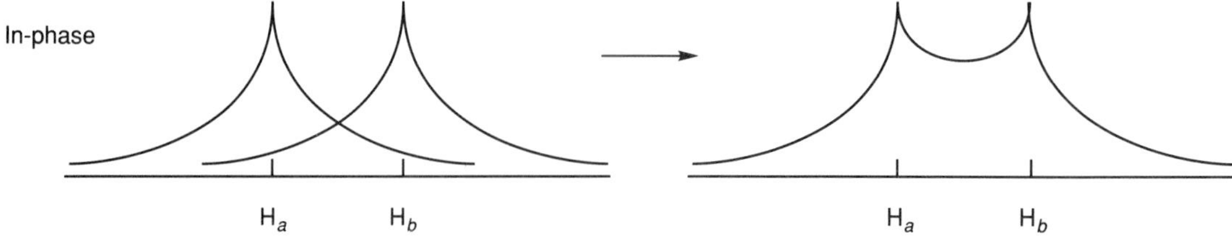

Fig. 3.4 Reinforcement of the waves from two hydrogen 1*s* orbitals on adjacent atoms (the 'in-phase' combination).

Here, as before, the symbol ψ represents the wave equation. The subscript 1*s* denotes the orbital name. The (H_a) and (H_b) labels denote two hydrogen atoms denoted *a* and *b*. The in-phase wave function combination has *cylindrical* symmetry. When an orbital possesses cylindrical symmetry it is referred to as a σ orbital.

σ: the Greek letter *sigma*.

Electron density is largely concentrated between the two nuclei for the in-phase combination (Fig. 3.1). The effect is to screen the positive nuclei from each other and for the negatively charged electrons to hold the nuclei together. Orbitals in which the electron density is so located are therefore referred to as *bonding molecular orbitals*.

$$\psi_{1s}(H_a) + \psi_{1s}(H_b) \rightarrow \text{in-phase combination} \tag{3.3}$$

3.5 Normalization constants

There is one further point to be made concerning the mixing of two (or more) orbitals. Mathematically there is no problem in simply adding together two wave functions, and then squaring to get a measure of electron density. Consider mixing a piece of red clay and a piece of yellow clay together. The result is an orange piece of clay, *but* one that is twice as big as the two component lumps. Something similar would happen on mixing two orbitals. The resulting orbital would be bigger than the component orbitals and in the case of orbitals this is not acceptable. It is therefore necessary to adjust the size of the orbital after mixing to make sure it is the expected size, that is, to ensure that when one electron is placed into the orbital, the total probability of finding the electron adds up to 1. Mathematically the resulting wave function is multiplied by a number less than one to achieve this result. Therefore the result of adding two hydrogen 1*s* wave functions together should be corrected by a multiplier, in this case *approximately* $1/\sqrt{2}$, to adjust the size of the orbital to the correct value.

Remember that obtaining values of electron density from a wave function involves squaring the wave function. Since the wave function of the in-phase combination is expressed by $\psi_{1s}(H_a) + \psi_{1s}(H_b)$, the square of this (Eqn 3.5) represents the electron density in the in-phase combination.

Normalization constants will be quoted where appropriate throughout the book, but usually without further comment. Later examples concern cases in which more than two wave functions are added together. In such cases the normalization constant will take some other value.

$$\psi_{\text{in phase}} = \frac{1}{\sqrt{2}}\left[\psi_{1s}(H_a) + \psi_{1s}(H_b)\right] \tag{3.4}$$

$$\left(\psi_{\text{in phase}}\right)^2 = \frac{1}{2}\left[\psi_{1s}(H_a) + \psi_{1s}(H_b)\right]^2 \tag{3.5}$$

3.6 Out-of-phase interaction of 1*s* orbital wave functions

The process of linear combination of orbitals necessarily generates the *same number* of molecular orbitals as the number of component atomic orbitals.

The point about the same number of resultant molecular orbitals arising from mixing the atomic orbitals is important. If starting with two atomic orbitals (such as the two 1*s* orbitals in H_2^+), then *two* molecular orbitals must result. So far only one of the resulting orbitals has been mentioned, the σ orbital which actually holds the electron in H_2^+. It is now time to consider the second molecular orbital of H_2^+.

The out-of-phase combination for the two hydrogen 1*s* orbitals is generated mathematically by subtracting (or multiplying the second wave function by –1 and adding the resultant wave function) the second wave function from the first at each point in space. The result of this operation is shown in Fig. 3.5 and in Eqn 3.6. The size of the *normalization constant* ($1/\sqrt{2}$) for the out-of-phase combination is the same as the in-phase case. Since squaring any negative number gives a positive number, although the wave function for the out-of-phase combination is negative in some regions, the electron density (Eqn 3.7) is always positive, as intuitively expected.

$$\psi_{\text{out of phase}} = \frac{1}{\sqrt{2}}\left[\psi_{1s}(\mathrm{H}_a) - \psi_{1s}(\mathrm{H}_b)\right] \tag{3.6}$$

$$\left(\psi_{\text{out of phase}}\right)^2 = \frac{1}{2}\left[\psi_{1s}(\mathrm{H}_a) - \psi_{1s}(\mathrm{H}_b)\right]^2 \tag{3.7}$$

Node: a surface at which the wave function is precisely zero.

The out-of-phase wave functions also has cylindrical symmetry and is therefore a σ orbital. If an electron is located in the out-of-phase combination orbital, the bulk of its density is situated remote from the internuclear region. The electron therefore does not screen the two nuclei from each other so effectively and the effect is for the two nuclei to see rather more of each other than their charges would prefer. A calculation reveals that the energy of an electron in the out-of-phase combination orbital is higher than the component atomic 1*s* orbitals. This kind of orbital is therefore referred to as an *antibonding molecular orbital*. To signify the significance of its antibonding nature the orbital is further labelled with an asterisk, and with a subscript to denote the nature of the contributor orbitals, making the full label σ_s* (spoken as *sigma star*).

Again, while the dot diagram (Fig. 3.6) is very descriptive, it is not easy

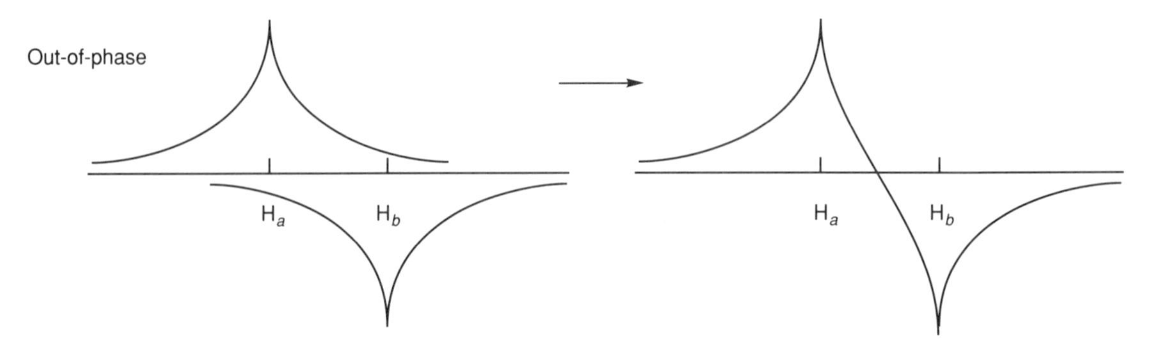

Fig. 3.5 Destructive interference of the waves from two hydrogen 1*s* orbitals on adjacent atoms (the 'out-of-phase' combination).

to construct routinely. A boundary diagram is the most convenient shorthand to represent these orbitals. Two representations are in common use and shown in Fig. 3.7. On the left, the plus and minus signs represent regions of space in which the wave function sign of the *molecular* orbital is positive and negative respectively. The shading of the two lobes in the representation on the right indicate that the wave function signs in each region are opposite. The dotted line represents the plane that bisects the H–H vector and for which the wave function takes the value of *precisely* zero. This plane is called a node.

The σ^* orbital in H_2^+ is unoccupied. The fact that an antibonding orbital contributes to the destabilization of a molecule does not mean that an antibonding orbital is never occupied. Far from it, as we shall see.

An empty orbital is not like an empty box. An empty box can be seen. So can the contents of the box when the box is filled. An orbital cannot be 'seen', whether it is empty or not. However, when an orbital is occupied, the *electron density* as given by the square of the wave function describing the orbital can be 'seen'.

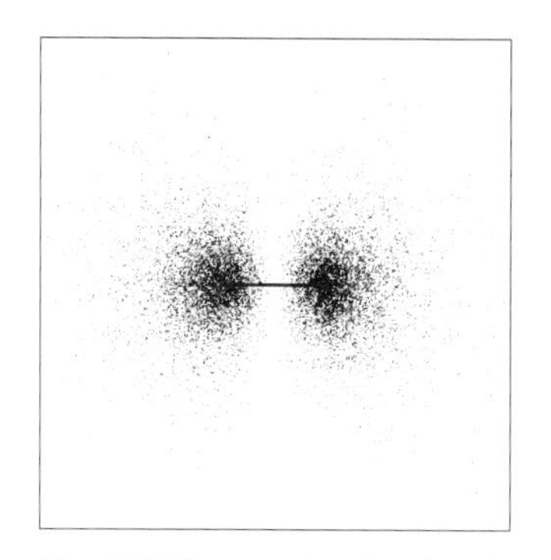

Fig. 3.6 Electron dot density plot of the σ^* orbital.

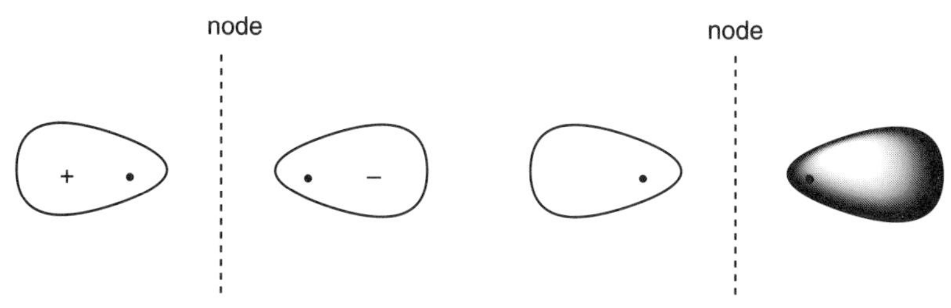

Fig. 3.7 Boundary representations of the antibonding orbital formed from two H 1*s* orbitals.

3.7 The H_2^+ energy level diagram

Without going into details here, it is possible to determine experimentally and to calculate the energy of the lowest energy orbital in the H_2^+ molecule. It was seen earlier that the energy of an electron in a given atomic orbital on hydrogen can be calculated. It turns out that the energy of the molecular orbital is lower than the energies of the hydrogen 1*s* orbital. Armed with this information, a schematic energy level diagram can be plotted. The bonding orbital is labelled σ to indicate its cylindrical symmetry.

The two levels on the left and right of Fig. 3.8 represent the 'energy level' of the 1*s* orbitals in the H atom and H^+ ion. The lower energy level in the centre represents the energy level of the lowest energy molecular orbital of H_2^+. The half arrows represent the electron of the H atom and of the H_2^+ molecule. The amount of energy ΔE by which the energy of the electron falls on going from atomic hydrogen to molecular H_2^+ represents the bonding energy of H_2^+. The calculated energy of the antibonding orbital is *higher* than the energy of the component 1*s* orbitals. This is therefore plotted above the component atomic energy levels. Note that this orbital is not occupied in H_2^+, but in related molecules (see below), it is occupied.

The Lewis structure representation of molecules has probably already instilled the idea that when *two* electrons hold two atoms together, the result is *one* bond. Two electrons constitute a bond. In H_2^+, there is a single electron holding the two nuclei together, so the bond order in H_2^+ is $^1/_2$.

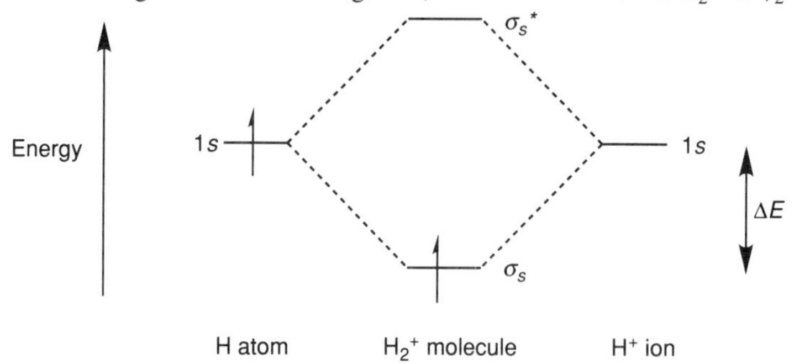

Fig. 3.8 An energy level diagram for H_2^+.

3.8 The H_2 energy level diagram

Once the ideas concerning the bonding in H_2^+ are grasped, representing the bonding for H_2 is quite straightforward. Again, just the H 1*s* orbitals are involved, so the possible orbital overlaps are the same as in H_2^+. The energy level framework of H_2^+ above is still applicable. The antibonding level is included for completeness, but again it is unoccupied.

The two electrons are both located in the lower energy level (Fig. 3.9), but spin paired, just as in the building-up principle for atoms discussed in Chapter 1. In fact, the various rules used for the building up principle for atoms (Hund's rule, Pauli exclusion principle, etc.) all operate for molecules in just the same way that they operate for atoms.

Since there are now two electrons in the σ-bonding orbital, the bond order of H_2 is said to be one.

The two molecules H_2 and H_2^+ behave very differently in a magnetic field. The molecule H_2 is *repelled* by a magnetic field. This is known as *diamagnetism*. The molecule H_2^+ is *attracted* by a magnetic field. This is known as *paramagnetism*. The origin of the paramagnetism is the unpaired electron in H_2^+. All molecules, except H_2^+, have an inherent diamagnetism but if there are unpaired electrons, the resulting paramagnetism outweighs the diamagnetism.

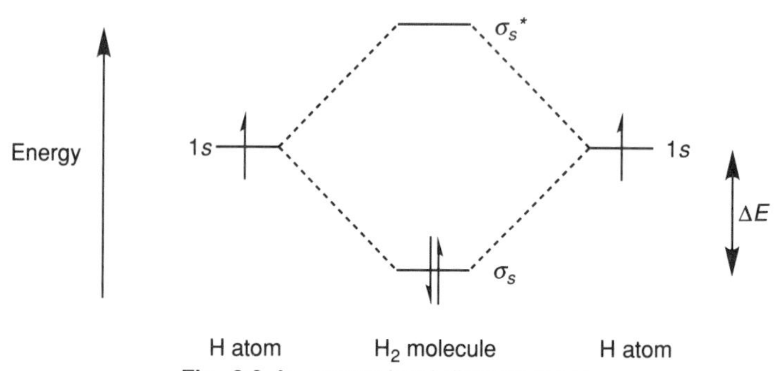

Fig. 3.9 An energy level diagram for H_2.

3.9 The H_2^- energy level diagram

The ion H_2^- is conceptually a combination of an H atom and an H^- ion. There are three electrons in the H_2^- ion. The first two occupy the σ-bonding orbital. Since there is no option, the third electron *must* occupy the σ^* orbital (Fig. 3.10). The two electrons in the σ-bonding orbital contribute one to the bond order. The third electron is regarded as contributing $-^1/_2$ to the bond order. This leaves a net bond order of $1 + (-^1/_2) = +^1/_2$, that is, the same as in H_2^+.

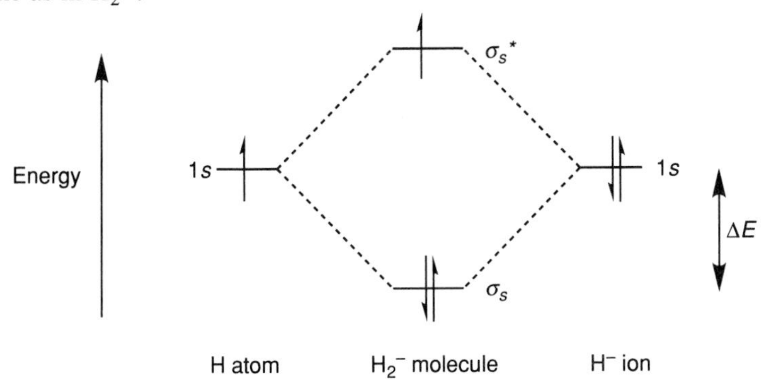

Fig. 3.10 The energy level diagram for H_2^-.

3.10 The He_2 energy level diagram

Helium is a monoatomic gas and He_2 is barely stable as a transient species. Each helium atom has two electrons to donate into the energy level diagram of He_2. The energy level diagram for He_2 is shown in Fig. 3.11.

The energy stabilization gained for the system by the two electrons in the bonding orbital is counterbalanced by the two electrons in the antibonding orbital. The net bond order is $+1 + (-1) = 0$. There is, therefore, no incentive, no energy gain, for the two helium atoms to bind together.

3.11 The mixing of *p* orbitals

A description of the bonding in diatomics such as O_2 requires some understanding of what happens to *p* orbitals on mixing.

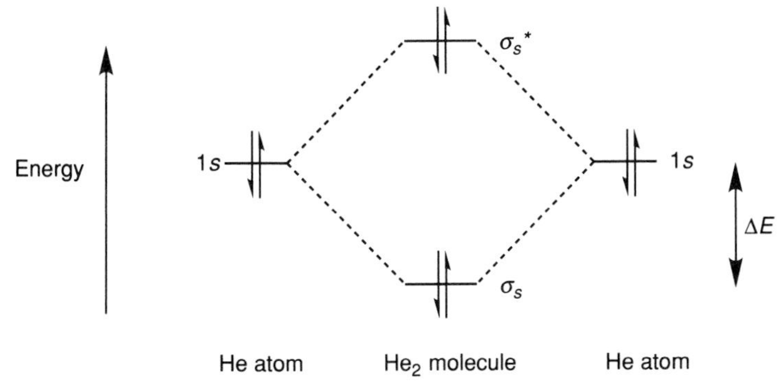

Fig. 3.11 The energy level diagram of He_2.

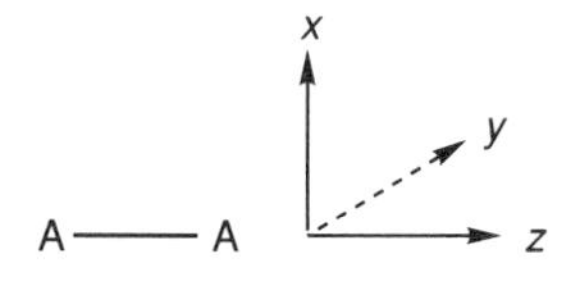

The first thing to sort out is which way the x-, y-, and z-axes are located. In practice, it doesn't matter too much but a convention helps to prevent misunderstandings. In a diatomic molecule (such as O_2), the z-axis is conventionally taken as the internuclear axis. In a diatomic molecule, the p orbitals are no longer equivalent. The p_z orbital is unique and different from the p_x and p_y orbitals.

p_z orbital mixing

When viewed along the intermolecular axis, a σ orbital of any type looks like a 's' orbital.

Just as two orbitals arise from the in- and out-of-phase mixing of two s orbitals, two arise from the mixing of two p_z orbitals (Fig. 3.12). The resulting orbitals possess cylindrical symmetry and are therefore classified as σ orbitals.

Because of the sign properties of the wave function in p orbitals, and because a universal z axis is used rather than local axes for each atom, note that the in-phase combination is derived by multiplying the p orbital on the second atom by –1 and adding the result to the p orbital on the first. If the atoms are identified as a and b then this is written as in Eqn 3.8.

The subscript p in the label σ_p simply refers to the fact that the orbital is made up from p orbitals. The σ_p orbital constitutes a bonding interaction since its calculated energy lies *below* that of the component p_z orbitals.

The out-of-phase combination (Eqn 3.9) is antibonding and labelled σ_p* since its calculated energy is above that of the component p_z orbitals. Note that, as for the σ bonding combination of H 1s orbitals in H_2, the electron density in the σ_p orbital is located between the two nuclei. Similarly, in the antibonding σ_p* orbital, the electron density is largely located away from the internuclear area.

$$\sigma_p = \frac{1}{\sqrt{2}}\left[p_z(\text{atom } a) + (-1) \times p_z(\text{atom } b)\right] \tag{3.8}$$

$$\sigma_p^* = \frac{1}{\sqrt{2}}\left[p_z(\text{atom } a) + p_z(\text{atom } b)\right] \tag{3.9}$$

p_x and p_y orbital mixing

The p_x orbital is set up to overlap with the p_x orbital of the second atom, but in a 'sideways' fashion (Fig. 3.13). The in-phase combination is depicted below by boundary representations.

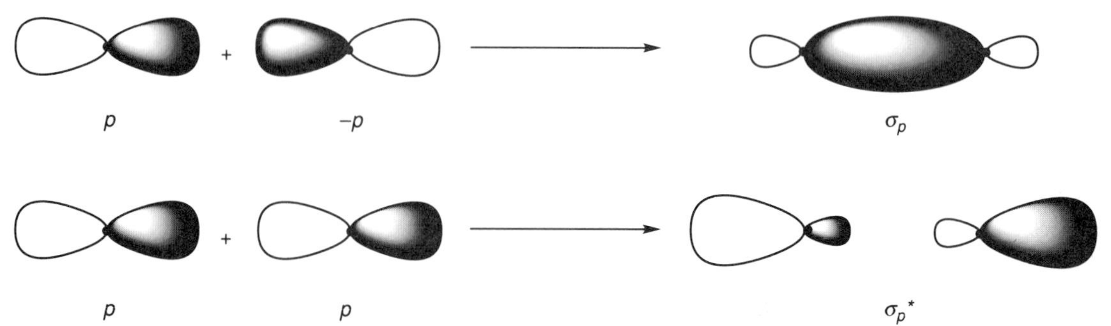

Fig. 3.12 The in-phase (top) and out-of-phase (bottom) mixing of $2p_z$ orbitals.

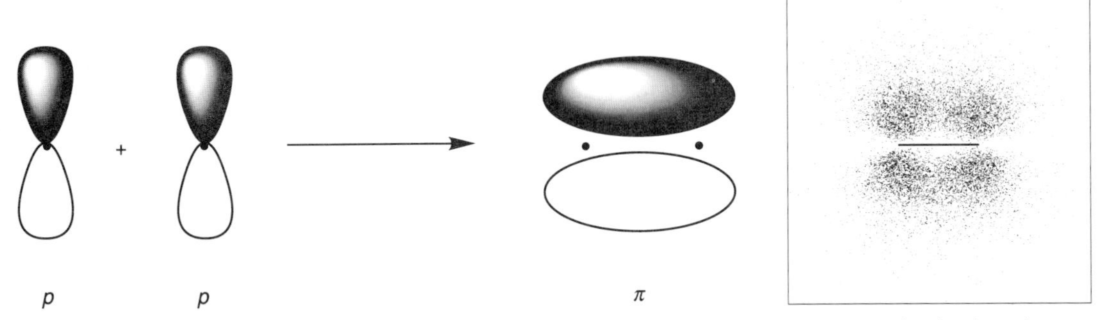

Fig. 3.13 The in-phase combination of two p_x orbitals and a dot representation of the electron density through a cross-section.

There are a number of points to be made. First, the symmetry of the orbital is clearly not cylindrical, so the label σ is not appropriate. Note that rotation about the intermolecular axis results in the wave function's positive lobe mapping on to the wave function's negative lobe. When an orbital shows this property it is referred to as a π orbital. It is bonding since its calculated energy is less than either of the component atomic orbitals. The in-phase combination is therefore written as in Eqn 3.10. The subscript p is included to denote explicitly its p orbital origins, but this is not obligatory as there is no source of confusion.

When viewed along the intermolecular axis, a π orbital looks like a '*p*' orbital.

$$\pi_p = p_x(\text{atom } a) + p_x(\text{atom } b) \qquad (3.10)$$

$$\pi_p^* = p_x(\text{atom } a) + (-1) \times p_x(\text{atom } b) \qquad (3.11)$$

The out-of-phase combination is shown in Fig 3.14. Again the resulting orbital has π symmetry, but it is antibonding since its calculated energy is greater than either of the component atomic orbitals. It is therefore denoted π^*. The out-of-phase combination is written as Eqn 3.11. Again the subscript p can be included to explicitly denote its p orbital origins, as in Eqn. 3.10, but again this is not obligatory.

In this type of bonding there is little electron density in the region of the internuclear axis between the nuclei in *either* the in-phase or the out-of-phase combinations. The bonding interaction in the π orbital comes from the two regions of electron density lying 'above' and 'below' the internuclear axis. The two lobes of the π interaction *together* constitute *one* orbital whereas the four lobes of the π^* interaction *together* constitute *one* orbital.

The situation for the p_y orbitals overlap is in appearance identical except that the π orbitals arising from p_y orbitals are at 90° to those of the p_x orbitals. Clearly by symmetry, the energies of the resulting π and π^* orbitals arising from p_y overlap are exactly the same as those arising from p_x overlap.

In π bonding the nuclei are relatively exposed to each other. In addition, the overlap of the π orbitals in this sideways fashion is less efficient than the end-to-end interaction of the p_z orbitals. The consequence is that π bonding is often less strong than σ_p bonding. Conversely, the destabilization of the π^*

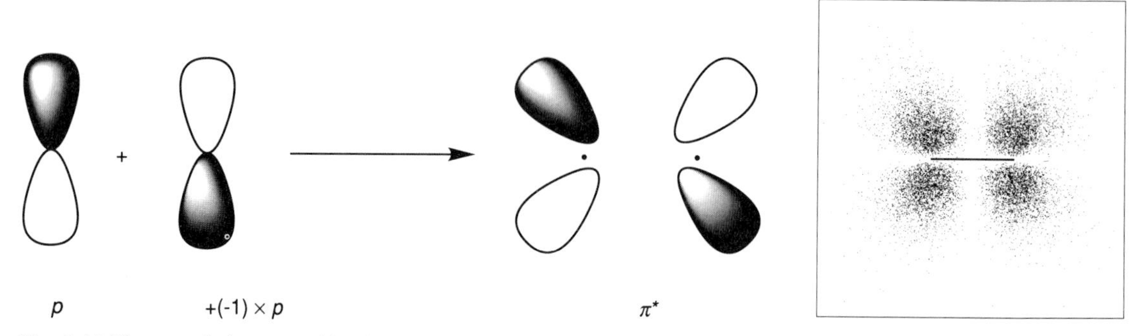

Fig. 3.14 The out-of-phase combination of two p_x orbitals and the electron dot density through a cross section of the orbital.

orbital is less than the σ_p* orbital. All of this is plotted on the energy level diagram of Fig 3.15.

The energy of the 2*s* orbital is below that of the 2*p* orbital, hence the ordering shown. The 2*s* orbital is lower in energy because the 2*p* orbitals are more shielded from the nucleus than the 2*s* orbitals (see Section 1.17). A consequence of the 2*s* orbital starting lower than the 2*p* orbital energies is that the σ_s* orbital is at a *lower* energy than the bonding σ_p bonding orbital. Just because an orbital is antibonding does not stop its energy being low and lying below the energy of bonding orbitals. The label antibonding indicates that the energy of that orbital is higher than those of *the component orbitals* of that antibonding orbital. If the component orbitals are themselves low in energy then the resulting antibonding orbitals may themselves be low in energy, and sometimes lower than other orbitals which are bonding relative to higher lying component orbitals.

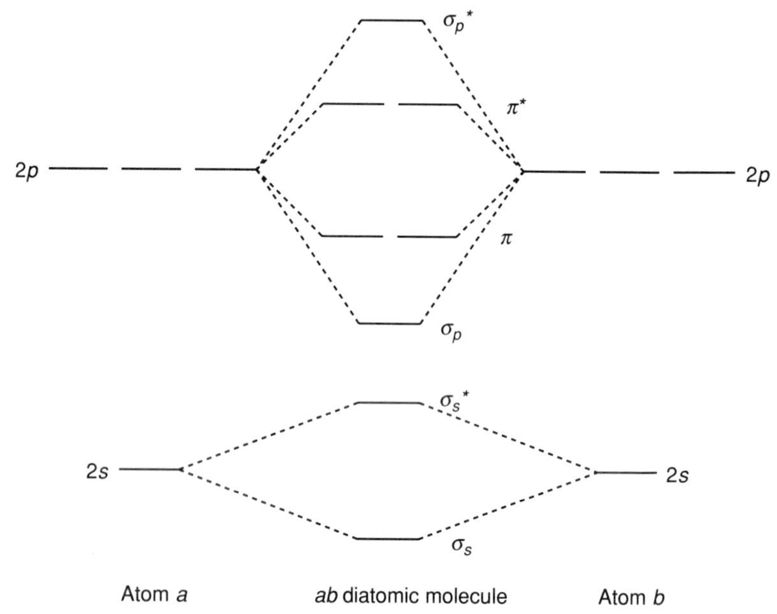

Fig. 3.15 An energy level diagram showing the relative positions of orbitals arising from overlap involving 2*s* and 2*p* orbitals (not to scale).

The σ_s* orbital will be populated before the σ_p-bonding orbital when applying the building-up principle to a diatomic molecule. When the π and π* orbitals are filled, they do so in a spin-parallel fashion, as required by Hund's rule of maximum multiplicity.

In the above description of π bonds, the two π orbitals arising from mixing of the p_x and p_y orbitals are said to be *degenerate* since their energies are *precisely* equal and required to be precisely equal by the symmetry of the molecule. Similarly, the two π* orbitals are degenerate.

3.12 Which orbitals can overlap?

It is not possible for any particular orbital to mix with just any other orbital. Certain criteria must be met. In particular the shape, and disposition of regions in space with different wave function signs is important. The *symmetry* of an orbital is said to be important. The concept of symmetry in a more formal discussion of bonding is extremely important and the reader should consult an appropriate book for a detailed discussion.

Some of the necessary concepts are made clear in a pictorial approach. In Fig. 3.16, two different ways for an *s* orbital to overlap with a *p* orbital are shown. The left hand side shows an 'end-on' overlap and the right a 'side-on' overlap. In the end-on interaction, the sign of the wave function is the same in all regions and so there is only reinforcement. There is a net bonding interaction. For the side-on overlap there are two regions of interest. On one side, a unshaded area overlaps with an unshaded area. The signs of the wave function in both these regions are the same and there will therefore be a degree of reinforcement in this region. However, on the other side an unshaded region overlaps with a shaded region. The wave function signs are different and therefore there will be degree of destruction in this region. From the symmetry of the diagram, it is clear that the degree of reinforcement on the one side is *precisely* countered by the degree of destruction on the other side. Because of the symmetry of the interaction, there is no net bonding or antibonding effect. Orbital overlaps for which some degree of mixing is expected and some for which there is not are shown in Figs. 3.17 and 3.18.

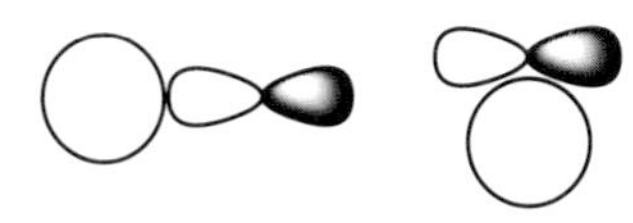

Fig. 3.16 Two ways for an *s* and a *p* orbital to interact.

3.13 Bonding in homodiatomic molecules

Armed with Figure 3.15, one can construct energy level diagrams for the first

 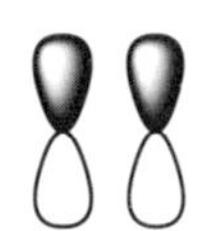 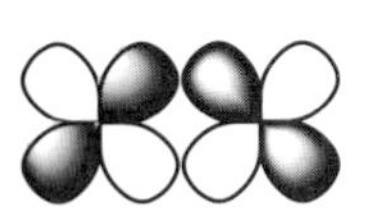 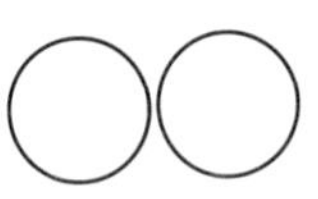

Fig. 3.17 Orbitals for which mixing is allowed by symmetry. Note that for each of these bonding combinations, there is a corresponding antibonding overlap.

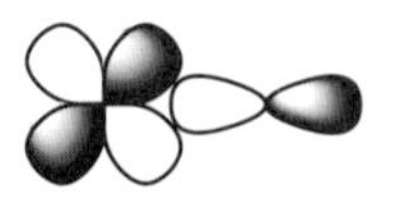 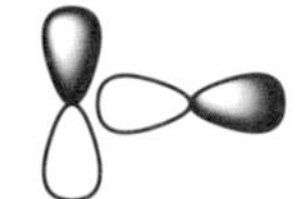  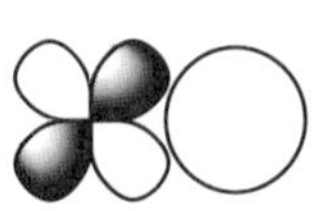

Fig. 3.18 Orbitals for which mixing is not allowed by symmetry.

row homodiatomic molecules Li_2 to Ne_2. It is best to start at the right-hand side of the periodic table with the hypothetical Ne_2 since there are complications for the earlier elements.

When considering the bonding of the diatomics Li_2 to Ne_2, strictly speaking, contributions from the 1*s* orbital overlaps should be included in the energy level diagrams, but they aren't. Why? As the nuclear charge on an atom builds up, the energy levels of all atomic orbitals drop. This is because the orbitals contract under the influence of the increasing nuclear charge. For the diatomics Li_2 to Ne_2, the 1*s* orbital contracts sufficiently far to make its size so small that there is *relatively little* effective overlap with the 1*s* orbital of the second atom. Thus electrons in the 1*s* orbitals play relatively little part in the bonding.

For most purposes, it is only ever necessary to consider the valence orbital electrons as playing any part in the bonding and all non-valence orbitals can be omitted from the diagram. This is indeed the basis for making the distinction between core and valence electrons. Of course there is nothing wrong with writing them in, but it is not *necessary* to do so.

Energy level diagram for Ne_2

A stable molecule Ne_2 does not exist. Neon is a monoatomic gas. It is instructive to build up the energy level diagram in order to see why neon is monoatomic rather than diatomic. The electronic configuration of Ne is $[He]2s^22p^6$ meaning that there are eight valence electrons, and 16 in Ne_2. These 16 electrons are fed in to the energy level diagram as in Fig. 3.19.

First, the σ_s orbital is bonding, but the bonding energy arising from its population is counteracted by equivalent energy increases experienced through population of the σ_s* orbitals. Occupancy of these two orbitals taken

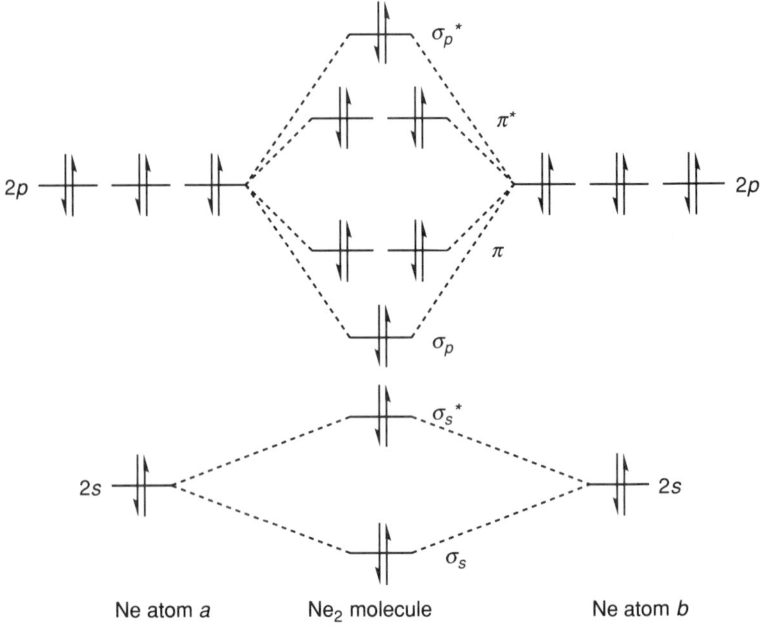

Fig. 3.19 Energy level diagram for Ne_2.

together make no net effect to the bonding. Secondly, for similar reasons, the four electrons of the two occupied π orbitals are counteracted by the those of the two occupied π^* orbitals. These eight electrons therefore make no net contribution to the bonding. Thirdly, the bonding effect arising from population of the σ_p orbitals is counteracted by population of the σ_p^* orbitals. The bond order is therefore 1 + (–1) + 2 + (–2) + 1 + (–1) = 0. The molecule does not exist, although a transiently stable Ne_2 species does exist through van der Waal interactions.

Energy level diagram for F_2

The electronic configuration of the fluorine atom is [He]$2s^2 2p^5$, a total of seven valence electrons. Fourteen electrons therefore populate the energy level diagram of F_2 (Fig. 3.20).

The bond order of F_2 is calculated as follows. Again the σ_s and σ_s^* electrons make no net contribution to the bonding. This is also true for the eight electrons that populate the π and π^* orbitals. This leaves just the two electrons in the σ_p orbital. These are *not* counteracted by any electrons in the unoccupied σ_p^* orbital. These two electrons alone are responsible for holding the two fluorine atoms together.

The bond order is therefore given by 1 + (–1) + 2 + (–2) + 1 = 1, which corresponds to the bond order of the Lewis structure. The structure predicted has no unpaired electrons, that is, F_2 is diamagnetic. The bond energy of F_2 is 155 kJ mol^{-1}, which is actually rather low, while the F—F bond length is 141.2 pm. An alternative way to arrive at the bond order is to apply formula 3.12, in which case the bond order is (8 – 6)/2 = 1.

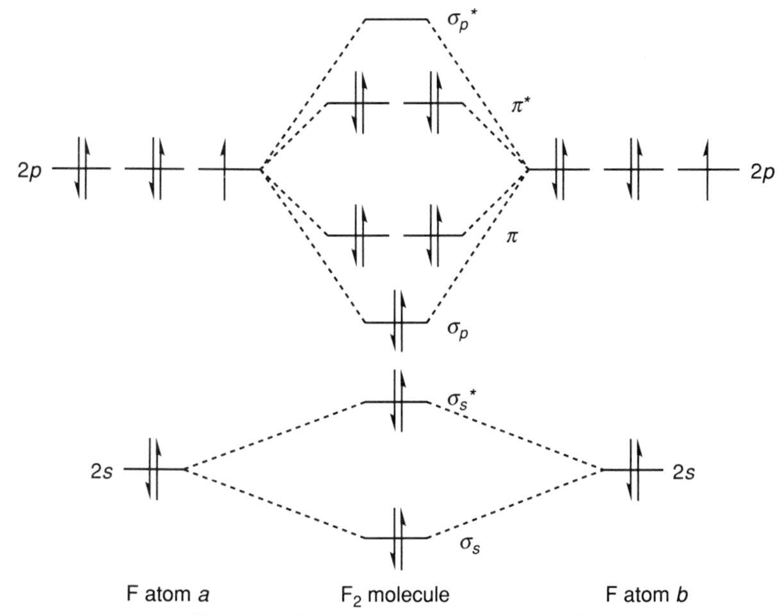

Fig. 3.20 The energy level diagram for F_2.

$$\text{bond order} = \frac{\text{no. bonding electrons - no. antibonding electrons}}{2} \qquad (3.12)$$

Energy level diagram for O_2

HOMO : the Highest Occupied Molecular Orbital.

LUMO : the Lowest Unoccupied Molecular Orbital.

Recall that the Lewis structure of O_2 predicts a double bond, but fails to account for its paramagnetism. If the molecular orbital energy level diagram representations are to be successful, then paramagnetism must be predicted.

The electronic configuration of the oxygen atom is [He]$2s^2 2p^4$. The same energy level diagram as given above is applicable for O_2, but there are now 12 electrons to feed into the diagram. Since there are two *fewer* electrons in the molecule relative to F_2, the π* orbitals are no longer fully occupied.

The first ten electrons feed in up to the π orbitals (Fig. 3.21). There are two more electrons to place and the next available place to put them is in the degenerate π* level. Hund's rule of maximum multiplicity requires that one electrons enters each of the two orbitals so that their spins are parallel. As a consequence, there are two unpaired electrons in the π* orbital. This energy level is the HOMO. The LUMO is the σ_p* orbital.

The bond order is calculated as follows. Again the contributions from the four electrons in the σ_s and σ_s* cancel and they make no net contribution to the bond order. The π orbital contribution counts as +2 from the four electrons in that degenerate level. The two electrons in the π* orbital contribute $2 \times -^1/_2 = -1$, since the two π* orbitals are each just half occupied. The σ_p orbital counts as +1 and there are no electrons in the σ_p* to counteract these bonding electrons. The bond order is therefore calculated as 1 + (–1) + 1 + 2 + $(2 \times -^1/_2)$ = 2. Therefore O_2 has a net bond order of +2.

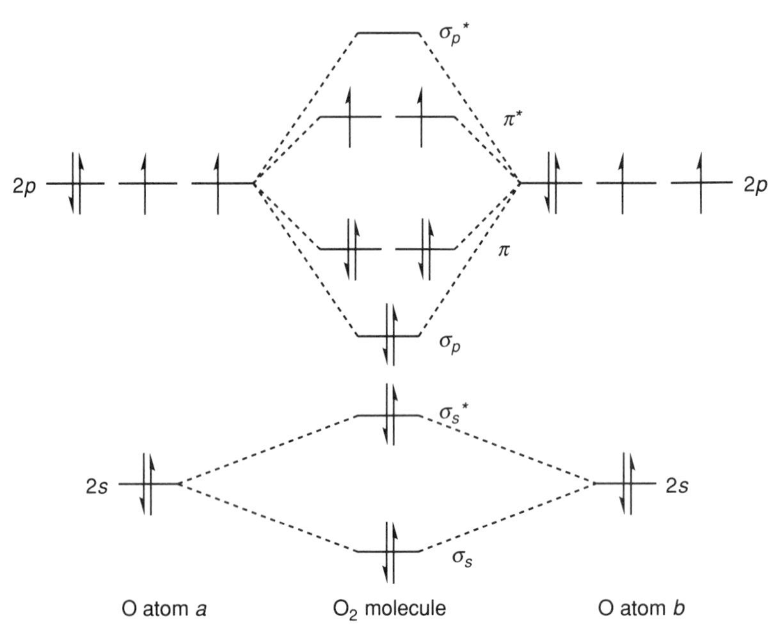

Fig. 3.21 Energy level diagram for O_2. The two unpaired electrons in the π* level are those responsible for the paramagnetism of O_2.

This again agrees with that suggested by the Lewis structure, but now the bonding in this MO description is very different to that of the Lewis model. The bond order is the same in the end, but the MO description successfully accounts for the paramagnetism of O_2, and also for the reactivity of O_2, which behaves as though it is a diradical (having two unpaired electrons).

The bond energy for O_2 is found experimentally to be 493 kJ mol^{-1} and the bond length is 120.7 pm. This is considerably stronger and shorter than F_2. It is interesting that removal of an electron from O_2 results in O_2^+ whose bond length is 111.5 pm and bond energy is 643 kJ mol^{-1}. Inspection of the energy level diagram reveals that this electron is necessarily removed from the π^* level. The bond order calculation for O_2^+ is $1 + (-1) + 1 + 2 + (-^1/_2) = 2^1/_2$. This correlates with a shorter bond length in O_2^+ relative to O_2 while the loss of an antibonding electron also helps to account for the increased bond energy in O_2^+ relative to O_2.

Energy level diagram for N_2

There is a complication at N_2. For the sake of simplicity, in all the energy level diagrams above, no indication was given of energy scale on the energy level diagram. Further, the fact that electrons in the constructed molecular orbitals *might interact with each other* was ignored. It was not necessary to include such interactions (although they are there) in the discussion of O_2 and F_2 since the bond orders and energy level orderings are not affected.

At N_2 the repulsive interactions between electrons in the various σ orbitals must be considered. These repulsions alter the shape of the orbital slightly from that when there are no interactions. Mathematically this is achieved by mixing in the mathematical description of other orbitals with the same symmetry.

Consider first the $2\sigma_s$ and $2\sigma_p$ orbitals, which are both symmetry-related (Fig. 3.22.) and fairly similar in energy. They therefore interact with each other. The fact that the orbitals are not identical in shape and energy does not matter. The mixing results in a bonding and an antibonding combination.

When two orbitals with dissimilar energies are allowed to interact, two new orbitals arise. The lower of the two new orbitals is bonding and the higher is antibonding. The lower of the new orbitals is closer in its properties to the lower of the component orbitals. The higher of the new orbitals is closer in its properties to the higher of the component orbitals.

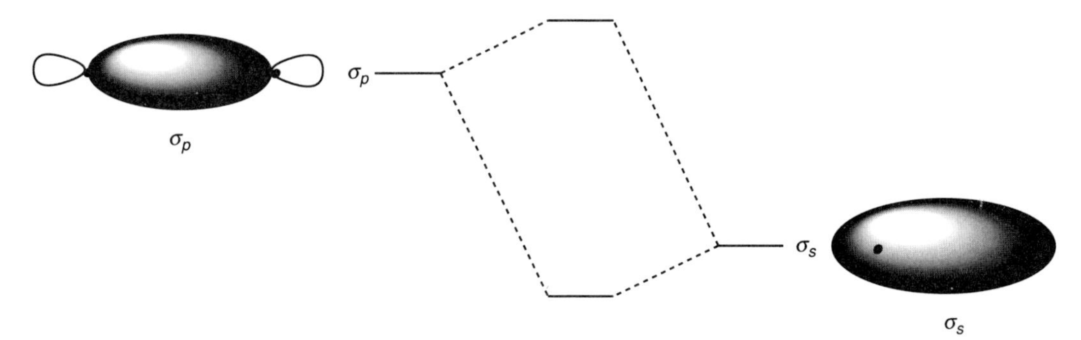

Fig. 3.22 The interaction (mixing) of σ_s and σ_p orbitals.

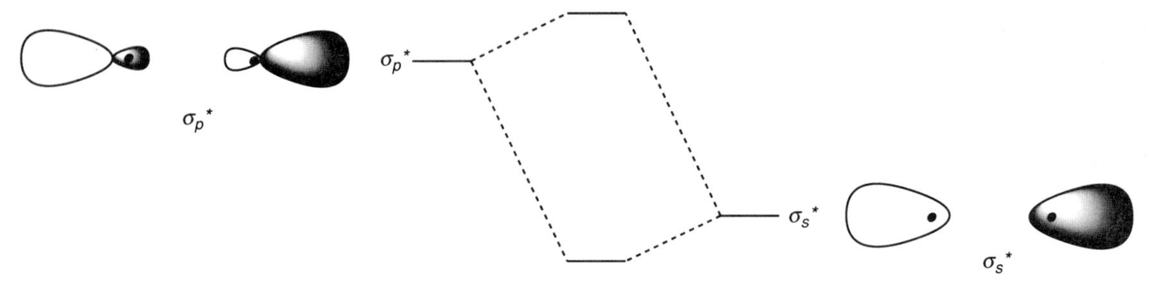

Fig. 3.23 The interaction (mixing) of σ_s^* and σ_p^* orbitals.

The situation for the σ_s and σ_p overlap is described by saying that the energy of the σ_s orbital is lowered, in the sense that the lower of the two new orbitals is closer in its properties to the σ_s than to the σ_p orbital. The lower of the two new orbitals is described as a modified σ_s orbital. Similarly the energy of the σ_p orbital is said to be raised through its modification by interaction with the σ_s orbital. Both of the new orbitals still possess cylindrical symmetry, so it is in order to describe them as σ orbitals. However, being a mixture of *s* and *p* orbitals, they are no longer quite σ_s and σ_p orbitals and so a numerical labelling system is used instead.

In a similar way, the σ_s* and σ_p* orbitals also possess the correct symmetry to interact with each other (Fig. 3.20). The effect is to modify and *lower* the σ_s* orbital and to modify and *raise* the energy of the σ_p* orbital.

There are no other orbitals of the correct symmetry for the π orbitals to overlap with, so any further interactions for them are safely ignored. The results of these interactions when plotted (Fig. 3.24) for N_2 show a *different*

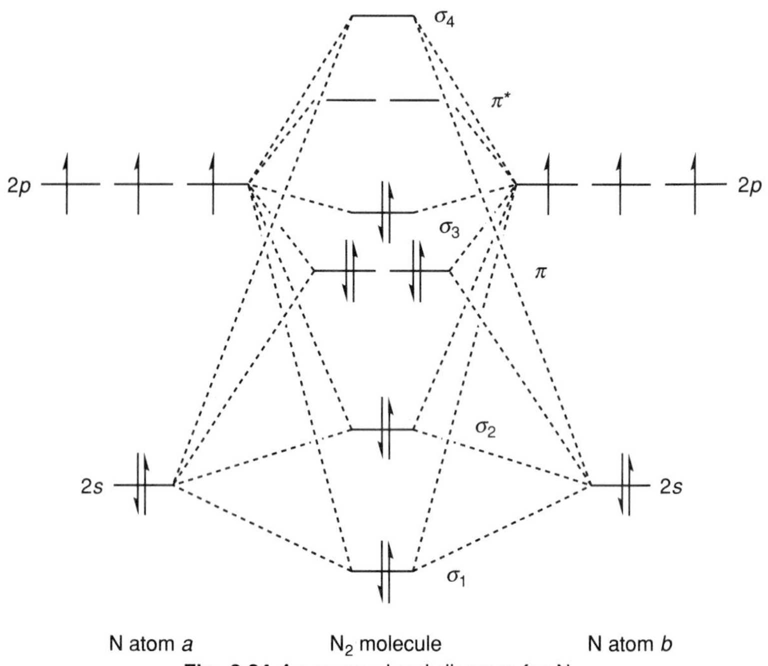

Fig. 3.24 An energy level diagram for N_2.

ordering for the energy levels in N_2 relative to those of Ne_2, F_2, and O_2.

Since the σ orbitals are no longer exactly σ_s, $\sigma_s{}^*$, σ_p, and $\sigma_p{}^*$, they are simply labelled σ_1, σ_2, σ_3, and σ_4. The important feature of this energy level diagram is that the modified σ_p orbital, σ_3, no longer lies *below* the π level. Instead it is moved *above* the π level. The consequence of this is that when following the building-up principle, the π orbitals fill *before* the σ_3 orbital.

The calculation of the bond order for N_2 is carried out in exactly the same way as for O_2 and F_2. The four electrons in the σ_1 and σ_2 orbitals correspond to the σ_s and $\sigma_s{}^*$ orbitals and more or less cancel each other out. The remaining six electrons occupy three bonding orbitals and are not counterbalanced by any electrons in their respective antibonding orbitals. The bond order calculation is therefore 1 + (–1) + 2 + 1 = 3. This is convenient since the bond order is the same as found in the Lewis structure.

Energy level diagram for C_2

The diatomic C_2 is not a species that can be put in a bottle, but it can be examined under special conditions. The electronic configuration of C is [He]$1s^2p^2$, so carbon possesses four valence electrons. The diatomic therefore has eight electrons with which to populate the energy level diagram. For this and all the remaining first row diatomics, the same energy level as used for N_2 is adequate (Fig. 3.25).

In this case, the σ_1 and σ_2 orbitals more or less cancel each other out, and it is permissible to say that they make no net contribution to bonding. There are four electrons in the π orbitals that are not counteracted by any electrons in the π^* orbital. The bond order calculation is therefore 1 + (–1) + 2 = 2. The net bond order is two. In this case the double bond is composed of two π

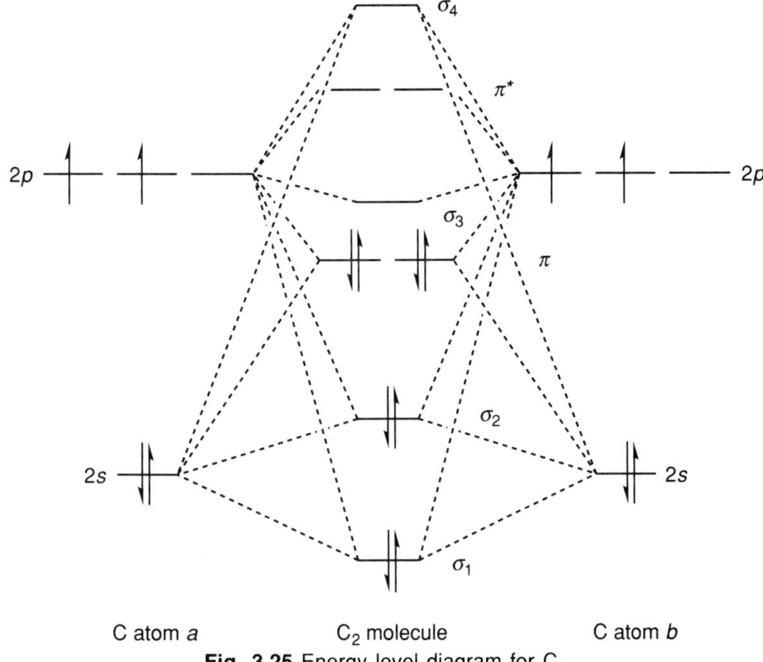

Fig. 3.25 Energy level diagram for C_2.

bonds unsupported by a σ bond. By comparison the double bond in ethene is made up from one σ and one π bond.

Energy level diagram for B_2

The diatomic B_2 is also not a species that one can put in a bottle, but it also can be detected and studied. Boron possesses three valence electrons as a consequence of its electronic structure $1s^2 2p^1$. There are therefore six electrons to place into the energy level diagram (Fig. 3.26).

Again the four electrons in the σ_1 and σ_2 orbitals largely cancel each other out for the purposes of calculating the bond order. The only bonding electrons are the two unpaired electrons in the π level. The net bond order is therefore $1 + (-1) + (2 \times {}^1/_2) = 1$. The bond order is 1, but in this case the bond is made up from two half-filled π orbitals.

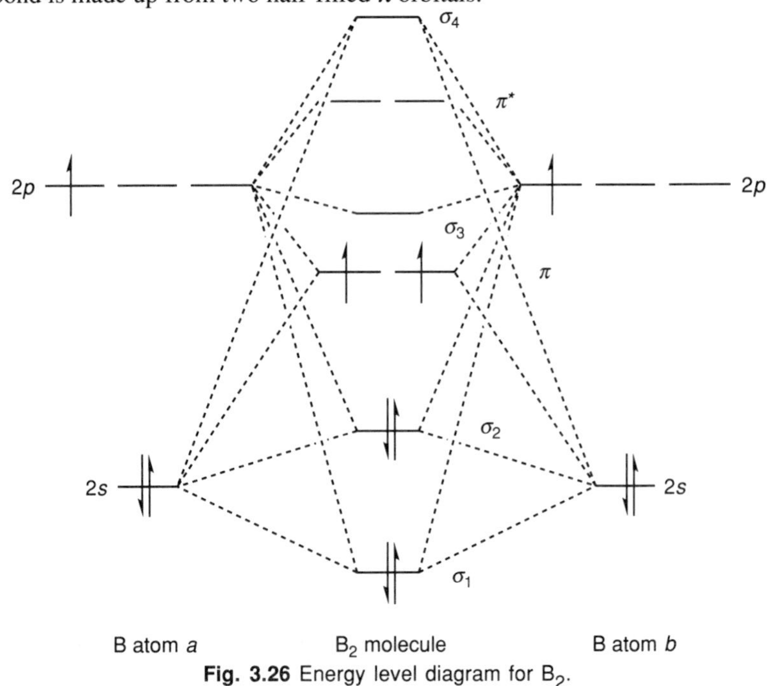

Fig. 3.26 Energy level diagram for B_2.

Bonding in Be_2

The diatomic Be_2 is definitely not an isolable species but it can be detected and studied under special conditions. Beryllium has the configuration [He]$1s^2$ and so there are four electrons to place into the energy level diagram (Fig. 3.27). These populate the σ_1 and σ_2 levels.

For Be_2 the bond order is therefore predicted to be zero, since the electrons in the σ_1 and σ_2 levels cancel each other out. However, because of the asymmetry in the diagram for the σ_1 and σ_2 levels, there is predicted to be a very slight bonding effect. However Be_2 has only been detected at very low temperatures and the bond energy is only about 4 kJ mol^{-1}, which is too weak for a 'normal' chemical bond.

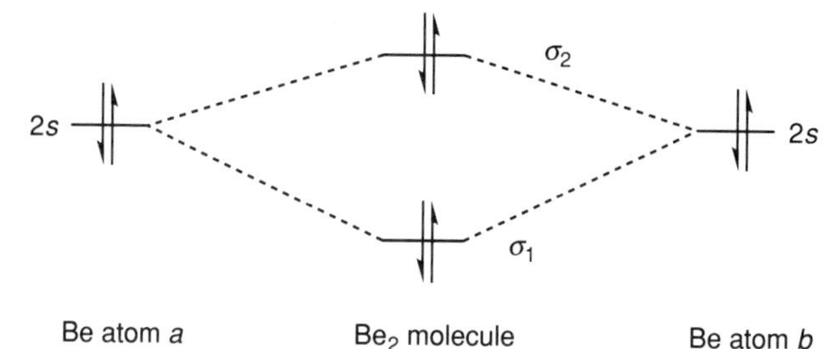

Fig. 3.27 Energy level diagram for Be_2.The *p* energy levels are omitted as they are empty.

Bonding in Li_2

The diatomic molecule Li_2 is known in the gas phase. It has just two electrons to place in the energy level diagram (Fig. 3.28) arising from the electronic configuration of [He]$1s^1$. These populate the σ_1 level and the bond order is therefore 1. The molecule is predicted to be diamagnetic, as is the case.

The bond energy for gas-phase Li_2 (101 kJ mol^{-1}) is considerably less than for H_2 (432 kJ mol^{-1}). Also the bond length in Li_2 (267.3 pm) is very much longer than in H_2 (74.1 pm). Part of the reason for the length of the Li—Li bond is the repulsion between the pairs of $1s$ electrons on each atom. Molecular orbital theory as outlined here shows that Li_2 is more stable than two Li atoms. In practice, when many lithium atoms come together, a more stable form of bonding is achieved in solid lithium metal.

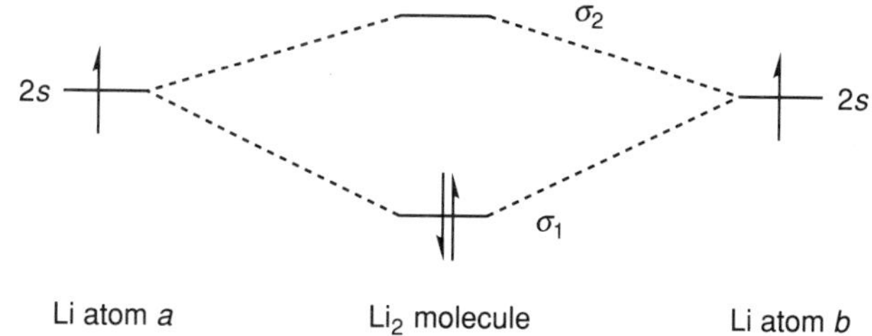

Fig. 3.28 Energy level diagram for Li_2. The *p* energy levels are omitted as they are empty.

3.14 Bond lengths, orders, and strengths

Bond lengths, orders, and strengths for the second period diatomics are tabulated in Table 3.1. For comparison the relevant data for He_2, H_2, and its ions are included.

There is no particular correlation of bond length with bond order or bond strength. There is something of a trend for bond strengths and bond orders. Higher bond orders tend to have higher bond strengths associated with them. The problem is that different bonds have different strengths, according to the efficiency of the orbital overlap, while overall bond energies are made up from a number of components (consider the situation in O_2, for instance).

Table 3.1. Bond orders, bond lengths, and bond energies for some homodiatomic molecules

Species	Bond order	Bond length/pm	Bond energy/kJ mol^{-1}
H_2^+	$^1/_2$	105.2	256
H_2	1	74.1	432
H_2^-	$^1/_2$	–	100–200
He_2	0	297	0.1*
Li_2	1	267.3	101
Be_2	0	–	4
B_2	1	159	289
C_2	2	124.25	599
N_2	3	109.8	942
O_2	2	120.7	493
O_2^+	$2^1/_2$	111.6	643
O_2^-	$1^1/_2$	135	395
O_2^{2-}	1	149	–
F_2	1	141.2	155
Ne_2	0	310	0.2*

*Van der Waal forces.

3.15 Valence shell orbital energies

An explanation as to why the energy level diagrams for Li_2 to N_2 show a different ordering of energy levels than O_2, F_2, and Ne_2 wopuld be useful. Why is it necessary to write the σ_3 level above the π level for the earlier elements? Examine Fig. 3.29, a graph of the orbital energies for the 2*s* and

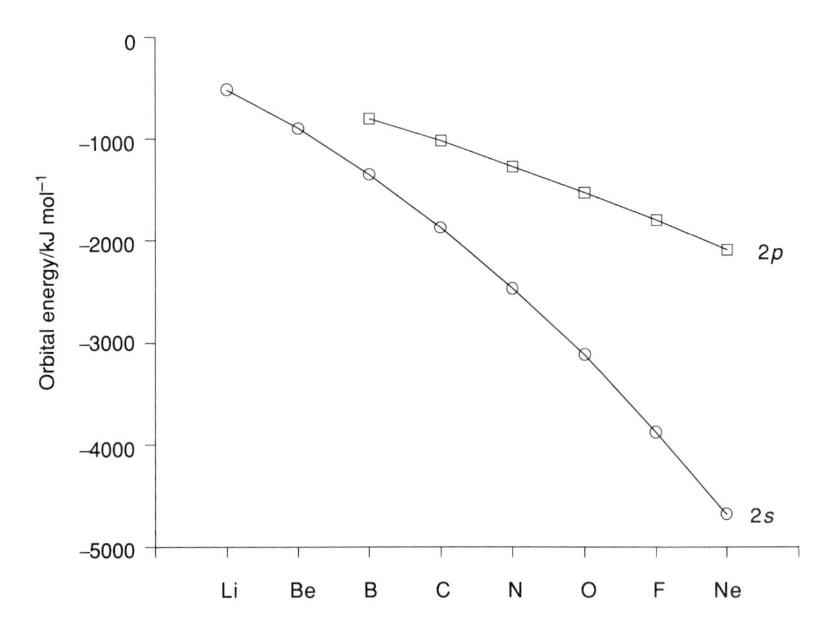

Fig. 3.29 Graph showing orbital energies of the first row elements.

$2p$ orbitals for the second period. The $2s$ and $2p$ energy levels start very close to each other on the left hand side of the periodic table and diverge towards the right. This is a consequence of the $2s$ level dropping more rapidly in energy than the $2p$ level (Section 1.17).

One of the factors that affects the extent to which energy levels mix is the closeness of their energies. If orbitals are close in energy, they mix to a larger extent. It is clear from Fig. 3.29 that the $2s$ and $2p$ levels are much closer together for the earlier elements. In fact they are not more than 1200 kJ mol^{-1} apart for the elements Li to N, but at least 1600 kJ mol^{-1} apart for O, F, and Ne. Because of these energy differences $2s$–$2p$ mixing is important for Li to N but can largely be ignored for O, F, and Ne.

3.16 Heterodiatomic molecules

The principles developed above for homodinuclear molecules also operate for heterodinuclear systems. The main difference is that orbitals which overlap to form bonding and antibonding combinations do not usually start at the same energy. This gives the energy level diagrams a skewed appearance.

When the two overlapping orbitals are corresponding orbitals of the same element, their energies are necessarily identical. In this situation (Fig. 3.30, left), the overlap is at its greatest and ΔE_{cov}, the measure of covalent bond energy is at a maximum. When one atom is more electronegative than the second, then the energy mismatch results in the orbital overlap being less efficient than in the homodinuclear case. In this case ΔE_{cov} is smaller.

This is not to say that the bonding between two dissimilar nuclei is weaker. Certainly the *covalent* bonding is less, but there is also an *ionic* contribution to the bonding. This may be very large. Consider a situation in which one of the atoms is *much* more ionic than the other. The orbital overlap diagram approaches the situation on the right of Fig. 3.30.

In a situation where dissimilar orbitals overlap, the bonding combination is more similar in appearance and energy to the lower energy contributing orbital, than to the higher energy contributing orbital. Corresponding to this, the antibonding combination is more similar in energy and appearance to the higher energy contributing orbital. This can be expressed differently by

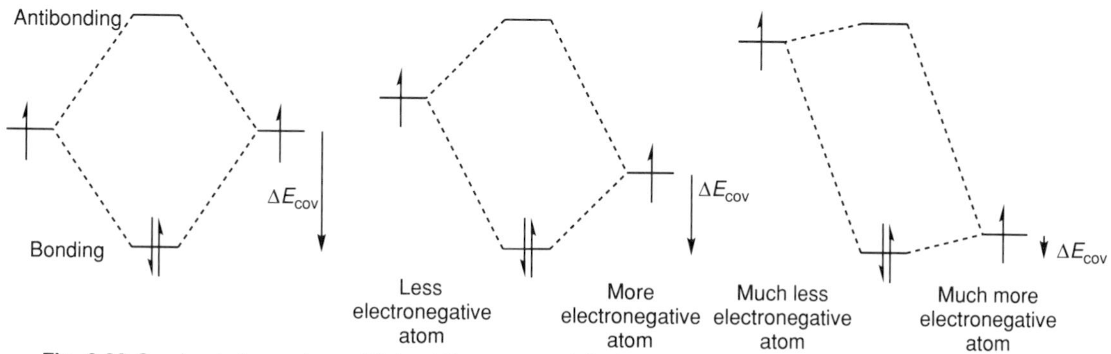

Fig. 3.30 Overlap between two orbitals of the same and dissimilar energies. In the centre, one atom is more electronegative than the second, while on the right, one atom is much more electronegative than the second.

saying that the bonding combination orbital is the lower energy contributing orbital, but somewhat modified. The antibonding combination therefore represents a modified orbital form of the higher energy contributing orbital.

In Fig. 3.30, for two atoms in which the electronegative difference is great, it should be clear that the bonding combination orbital must therefore be, in essence, a very slightly modified atomic orbital from the more electronegative atom. If there is one electron in each of the contributing atomic orbitals before mixing, an electron is in effect transferred from the more electropositive atom to the more electronegative atom. The situation approaches the formation of an ion pair in which there is a very small covalent contribution to the overall bonding.

The total bonding is expressed as a combination of a covalent contribution and an ionic contribution.

Total bonding energy = covalent contribution + ionic contribution

In many cases the ionic contribution is very large. A situation in which this is the case would be the formation of an ion pair from sodium and chlorine.

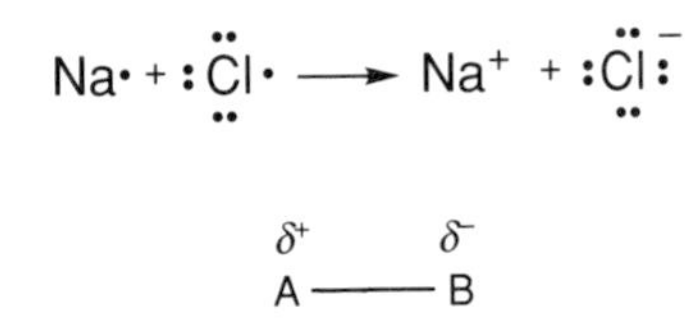

The Lewis representation shows the complete transfer of an electron from sodium to chlorine. The molecular orbital diagram (Fig. 3.30, right) would suggest that there is at least a small covalent contribution to the bonding.

If the difference in electronegativities of the two component atoms is less extreme then the covalent contribution to the total bond is larger and from the ionic contribution smaller. There is a 'spectrum' of bond natures with all real bonding interactions lying between the two extremes of the purely covalent and the purely ionic.

Hydrogen fluoride

In order to construct an energy level diagram for HF it is necessary to have some idea as to the relative energies of the valence orbitals, 1*s* for hydrogen and 2*s* and 2*p* for fluorine. These are indicated approximately in Fig. 3.31.

The 1*s* hydrogen orbital is well above the fluorine 2*s* orbital and to a lesser extent above the 2*p* level. The effect of the very large difference in energies between the hydrogen 1*s* orbital and the fluorine 2*s* orbital is that there is approximately no interaction. This is despite the fact that two *s* type orbitals are permitted to overlap by their symmetry. However by symmetry, and because of the lesser energy difference, the hydrogen 1*s* orbital does mix with the fluorine $2p_z$ orbital. The result is a bonding and an antibonding level. By symmetry, the fluorine $2p_x$ and $2p_y$ orbitals are unable to interact with the hydrogen 1*s* orbital.

According to the earlier discussion, the σ^* orbital is largely an atomic hydrogen 1*s* orbital. The σ-bonding orbital is largely a fluorine $2p_z$ orbital. In effect there is considerable transfer of electron density from hydrogen atom to fluorine. This results in a charge imbalance in the molecule, with the hydrogen end somewhat positive and the fluorine end somewhat negative. The 2*s*, $2p_x$, and $2p_y$ orbitals are all non-bonding, the 2*s* orbital because of an energy mismatch and the $2p_x$ and $2p_y$ orbitals because there are no symmetry related orbitals with which they may interact. Note they are unused in Fig. 3.31.

δ^+ δ^- H—F ≡ H—F (with ⟼ arrow above)

Two ways to denote that the fluorine atom carries a partial negative charge and hydrogen a partial positive charge in HF.

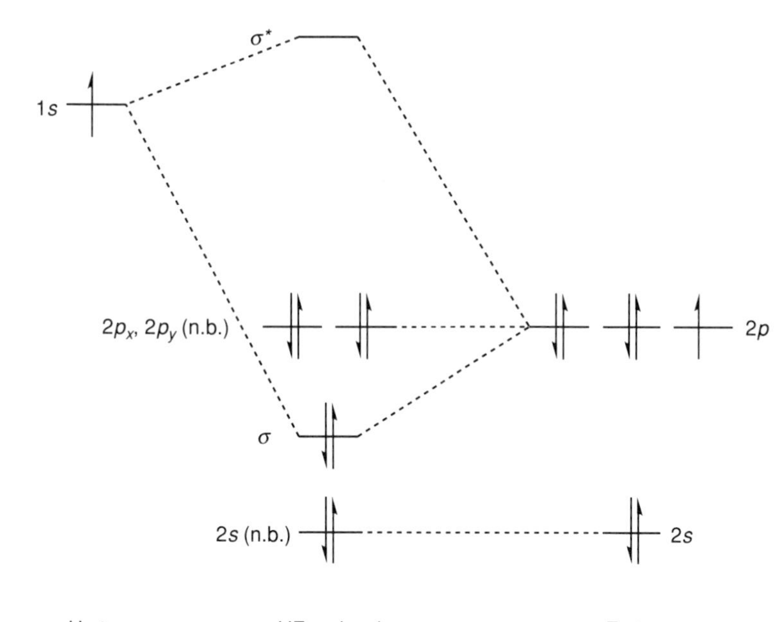

Fig. 3.31 Energy level diagram for HF. (n.b. = non-bonding)

In constructing Fig. 3.31 some liberties were taken. In reality, some small contribution should be considered from the fluorine 2*s* orbital, perhaps by slightly mixing it with the fluorine $2p_z$ orbital (hybridization). Also, the energy of the non-bonding π orbitals (fluorine $2p_x$ and $2p_y$ atomic orbitals) in the HF need not lie *precisely* at the same level as the atomic orbitals in fluorine, after all, their environment is inside a *molecule* rather than an *atom*.

Carbon monoxide, CN^-, and NO^+

Carbon, $[He]^2s^22p^2$, possesses four valence electrons. Oxygen, $[He]2s^22p^4$ possesses six valence electrons. An energy level diagram for carbon monoxide should therefore account for the placing of these ten valence electrons. Both carbon and oxygen have populated *s* and *p* orbitals and it is therefore appropriate to adapt the energy level diagrams used above for homodinuclear diatomics. The form of the energy level diagram is shown in Fig. 3.32.

Note that the oxygen 2*s* and 2*p* orbitals are lower in energy than the carbon 2*s* and 2*p* orbitals. This is because of the greater effective nuclear charge for oxygen. The form of the diagram is therefore slightly asymmetric. The ordering of the energy levels in the diagram follows the same ordering as in the diagram for N_2, which is, of course, also a 10-electron system. This correspondence of electronic configuration means that N_2 and CO are said to be *isoelectronic*. Both these two molecules are isoelectronic with other diatomics, notably CN^- and NO^+. Therefore the same ordering in the energy level diagram is applicable for these two molecules, the only differences being in the absolute values for the various energy levels.

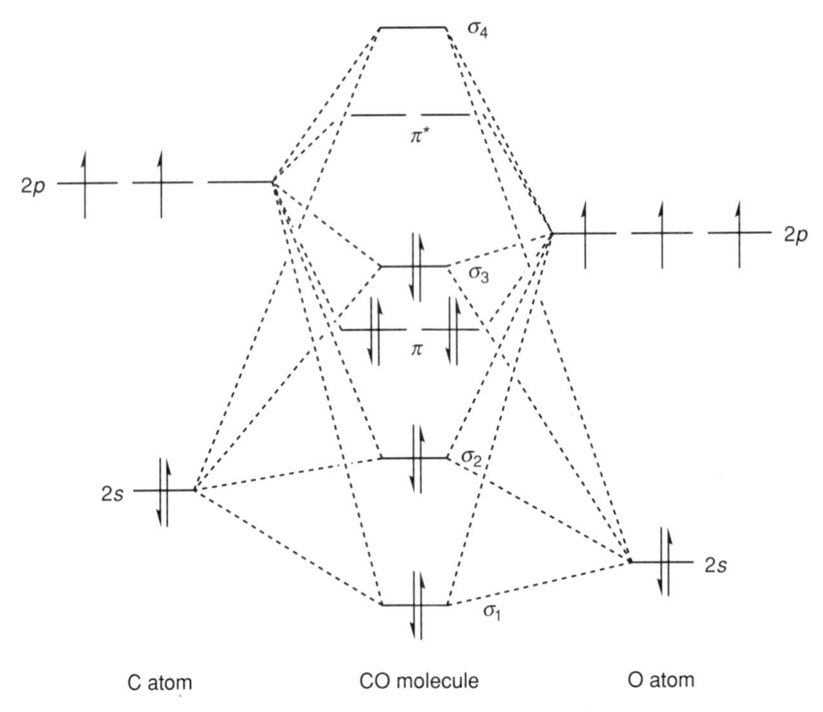

Fig. 3.32 Energy level diagram for CO.

A closer examination of the bonding situation for CO will reveal that σ_3 corresponds to an orbital that is largely a carbon-based lone pair. The σ_2 level is largely an oxygen lone pair with σ_1 being the carbon–oxygen σ bond. The π^* orbitals are unoccupied. From the point of view of the coordination chemistry of CO, the filled σ_3 and empty π^* orbitals are critical since they are involved in bonding interactions with various transition metal orbitals.

3.17 Exercises and problems

1. The cation He_2^+ has a bond length of 108 pm. Write down an energy level diagram for this species and determine the bond order.
2. Describe the bonding in LiH based upon a molecular orbital model.
3. Describe the bonding in OH^- and $OH^\bullet$ based upon a molecular orbital model.
4. Write out energy level diagrams for the species O_2^+, O_2^-, and O_2^{2-}. In each case determine the bond order.
5. For which of the diatomic species Li_2 to Ne_2 do the corresponding monocations and monoanions have higher bond orders than the corresponding neutral species?

4 Molecular geometry: VSEPR

One of the more obviously important properties of any molecule is its shape. Clearly it is very important to know the shape of a molecule if one is to understand its reactions. It is also desirable to have a *simple* method to *predict* the geometries of compounds. For main group compounds, the VSEPR method is such a predictive tool and unsurpassed as a handy predictive method. It is a remarkably simple device that utilizes a simple set of electron accounting rules in order to predict the shape of, in particular, main group compounds. Organic molecules are treated just as successfully as inorganic molecules. The assumptions and simplifications required by the method as outlined here should not encroach too far into descriptions of bonding: it is enough that the shape of molecules are successfully predicted.

VSEPR: Valence Shell Electron Pair Repulsion.

Application of the VSEPR method requires some simplifying assumptions about the nature of the bonding. Despite this, the correct geometry is nearly always predicted, and the exceptions are often rather special cases. In a complete analysis of the geometry of a molecule it would be necessary to consider such factors as nuclear–nuclear interactions, nuclear–electron interactions, and electron-electron interactions. In the VSEPR method outlined here it is *assumed* that the geometry of a molecule depends *only* upon electron-electron interactions.

Some information is also needed before one can apply successfully the VSEPR rules. The connectivity of the atoms in the molecule must be known; in other words, one needs to know which atoms are bonded together. For instance, does the empirical formula C_2H_6O refer to MeOMe or EtOH? It is also necessary to write down a Lewis electron dot structure for the molecule, but it will emerge that one can sometimes use a simplified Lewis structure.

4.1 Assumptions about the nature of the bonding

The underlying assumptions made by the VSEPR method are the following.

- Atoms in a molecule are bound together by electron pairs. These are called *bonding pairs*. More than one set of bonding pairs of electrons may bind any two atoms together (multiple bonding).
- Some atoms in a molecule may also possess pairs of electrons not involved in bonding. These are called *lone pairs* or *non-bonded pairs*.
- The bonding pairs and lone pairs around any particular atom in a molecule adopt positions in which their mutual interactions are minimized. The logic here is simple. Electron pairs are negatively charged and will get *as far apart from each other as possible*.
- Lone pairs occupy more space than bonding electron pairs.
- Double bonds occupy more space than single bonds.

Table 4.1. VSEPR geometries

Electron pairs	**Geometry**
2	linear
3	trigonal planar
4	tetrahedral
5	trigonal bipyramidal
6	octahedral

It is necessary to know the most favourable arrangement for any given number of electron pairs surrounding any particular atom. These arrangements are found using simple geometrical constructions. This involves placing the nucleus of the atom in question at the centre of a sphere and then placing the electron pairs on the surface of the sphere so that they are as far apart as possible. The resulting arrangements are often intuitively obvious.

For the case of just two electron pairs (Fig. 4.1) the arrangement is simple and the minimum energy configuration is when the electron pairs form a linear arrangement with the nucleus. In this configuration the electron pair–nucleus–electron pair angle is 180°. The coordination geometry of the central atom is described as *linear*.

Three electron pairs arrange themselves *trigonally*, that is with bond angles of 120°. For four electron pairs, one might expect the square-planar geometry to be favourable. However, tetrahedral bond angles are 109.5°, larger than the square-planar angles of 90°. If, for the purposes of illustration, the electron pairs are assumed to be points then it should be clear that, for similar bond lengths, the electron pairs are necessarily further apart in the tetrahedral arrangement than in a square-planar arrangement. There is more electron pair–electron pair repulsion in the square-planar geometry and so the *tetrahedral* geometry is favoured.

The case of five coordination is a little trickier. Most molecules whose shape is determined by five electron pairs are *trigonal bipyramidal*. There are two environments in a trigonal bipyramid, axial and equatorial. These two environments are chemically distinct. There is another very reasonable

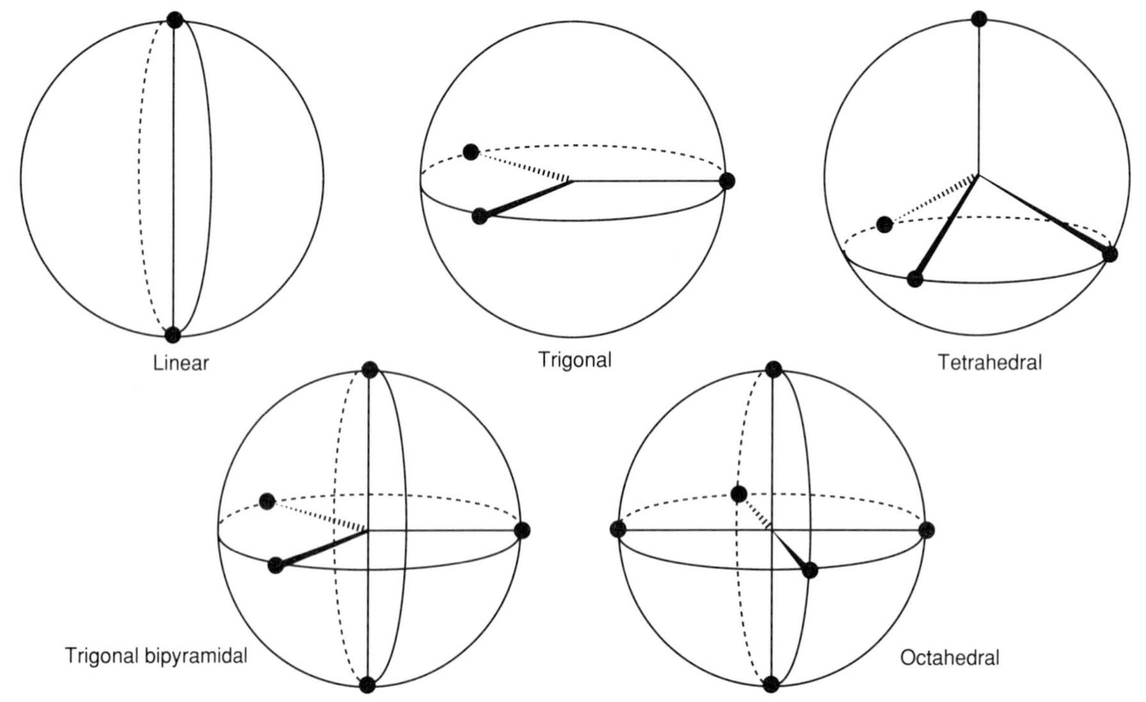

Fig. 4.1 Arrangements for 2- to 6-electron pairs on the surface of a sphere.

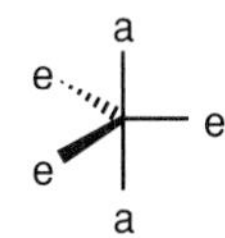

candidate, and that is the square-based pyramid. In effect this arrangement is an octahedron in which one group is removed and in which the four adjacent groups move down slightly to occupy partially the resulting vacancy. In practice, this geometry is only a little disfavoured relative to the trigonal bipyramid and the square-based pyramidal geometry is very important in the interconversion of axial and equatorial environments in trigonal bipyramids.

For six-coordinate systems, the *octahedral* geometry is by far the most important. An alternative geometry, the trigonal prism is uncommon.

A Lewis single bond is always regarded as a σ bond. Recall that Lewis dot structures allow double and triple bonds between some elements. For VSEPR calculation purposes, a double bond is always regarded as a $\sigma+\pi$ bond interaction, while a triple bond is always treated as a $\sigma+2\pi$ bond interaction. This simplification is permissible despite the MO treatments of Chapter 3 showing that the nature of some multiple bonds is somewhat different. However, double and triple bonds still *only occupy one coordination vertex*. The shape of a molecule is therefore dictated by the σ bond framework. Each vertex of the coordination polyhedron is necessarily occupied by a σ bond (possibly supported by π bonds) or a lone pair of σ symmetry, otherwise the vertex would not exist. The determination of a molecule's geometry therefore reduces to a calculation of the number of electrons contained in the σ orbitals. Electrons in π bonds must therefore be recognized and care taken not to confuse them with σ bond electrons. The following procedure, which is not unique, works well.

1. Draw a Lewis structure based upon the atom of interest. In the following examples it will become clear that some simplifications are possible.
2. Determine the number of valence electrons on the central atom.
3. Write down a modified Lewis structure by assigning all atoms or groups bonded to the atom in question as *singly*, *doubly*, or *triply* bonded. It is not necessary to write in lone pairs, the calculation will determine these. Assign all singly bonded groups as shared electron pair bond types, with the exceptions of dative bound groups discussed below. Always regard groups such as =O and =S as double bonded to the central atom with the double bond consisting of a σ and a π bond, both of which are electron pair bonds. Regard groups such as ≡N and ≡P as always triple bonded, with the triple bond consisting of a σ and two π bonds, all of which are shared electron pair bonds.
4. Groups for which the octet rule is satisfied, such as NH_3, are regarded as lone pair electron donors to an atom such as B in $H_3N{\rightarrow}BF_3$. That is, there is still a σ bond, but both electrons originate from the attached group (N). These are *dative* bonded groups. Conversely if N is under consideration as the central atom in $H_3N{\rightarrow}BF_3$, then the BF_3 group is bonded through a single bond to the central atom (N), but here both electrons of the bond originate from the central atom. In this case, there is *no net contribution* of electrons from the attached group to the central atom.

Table 4.2. Classification of formal bonding type to central atom

Single	Double	Triple
F, Cl, Br, I		
OH, SH	=O, =S	
NH_2	=NH, =PH	≡N, ≡P
Me, Ph	$=CH_2$	≡CH
H		
$SiMe_3$		

5. The coordination geometry is dictated by the σ framework only. It is now necessary to discount those central atom electrons that are involved in π bonds. Since each π bond is a shared electron pair with one electron arising from each atom, subtract one electron for each π bond involving the central atom.
6. Any overall charge on the molecule is *always* assigned to the central atom, even if later reflection requires that it may be best to assign it elsewhere. Thus, a negative charge constitutes an additional electron for the central atom, while a positive charge requires subtraction of one electron from the central atom electron count.
7. Divide the total number of electrons associated with the σ framework by 2 to give the number of σ electron pairs. Assign a coordination geometry and perhaps distinguish between isomers.

A statement of rules always seems more imposing than their actual operation. It is therefore appropriate to examine some sample calculations. The calculation for methane shows that the carbon atom is associated with 8 electrons in the σ framework. This corresponds to four shape-determining electron pairs. The coordination geometry of carbon is consequently tetrahedral. There are four bonded groups, therefore there are no lone pairs.

Ammonia also has four electron pairs and the *coordination geometry* of nitrogen is based upon a tetrahedral arrangement of electron pairs. There are

The geometry of methane, CH_4

Methane, CH_4

Lewis structure:

```
    H
    |
H — C — H
    |
    H
```

central atom: **carbon**	
valence electrons on central atom:	4
4 H each contribute 1 electron:	4
total:	8
divide by 2 to give electron pairs:	4
4 electron pairs: *tetrahedral* geometry	

The geometry of ammonia, NH_3

Ammonia, NH_3

Lewis structure:

```
H — N — H
    |
    H
```

central atom: **nitrogen**	
valence electrons on central atom:	5
3 H each contribute 1 electron:	3
total:	8
divide by 2 to give electron pairs:	4
4 electron pairs: *tetrahedral* arrangement	

three bonded groups, therefore there is one lone pair. However since the lone pairs are 'invisible', the *shape* of ammonia is *pyramidal.*

Consider a bonding pair of electrons. The two electrons are located between two nuclei, and are attracted by both. A lone pair is different. It is necessarily only attracted to one nucleus and the consequence is that it adopts a position effectively *closer* to that one nucleus than the bonding pairs of electrons. This means that the effective solid angle occupied by a lone pair is *greater* than that occupied by a bond pair. Lone pairs demand greater angular room, and are located closer to their atoms than bond pairs. The consequence of this for ammonia is that the lone pair makes room for itself by pushing the three hydrogen atoms together a little and the H—N—H bond angles are slightly less (106.6°) than the ideal tetrahedral angle of 109.5°.

Water has four electron pairs and the *coordination geometry* of oxygen is based upon a tetrahedral arrangement of electron pairs. Since there are only two bonded groups, there are two lone pairs. Since the lone pairs are not seen, the *shape* of water is bent. The two lone pairs compress the H—O—H bond angle below the ideal tetrahedral angle to 104.5°.

Boron trifluoride only has six valence electrons and is one of the relatively rare second period covalent molecules that disobeys the octet rule. There are three bonded groups and so no lone pairs. Six electrons implies three electron pairs and therefore a *trigonal* geometry. Since it is deficient two electrons from the octet, a characteristic reaction is to react with lone pair donors. Acceptance of a pair of electrons makes BF_3 a *Lewis acid*. The electron pair donor, say, NH_3, is a *Lewis base*. The product of the reaction between NH_3

Water, OH_2

Lewis structure: H—O—H

central atom: **oxygen**	
valence electrons on central atom:	6
2 H each contribute 1 electron:	2
total:	8
divide by 2 to give electron pairs:	4
4 electron pairs: *tetrahedral* arrangement	

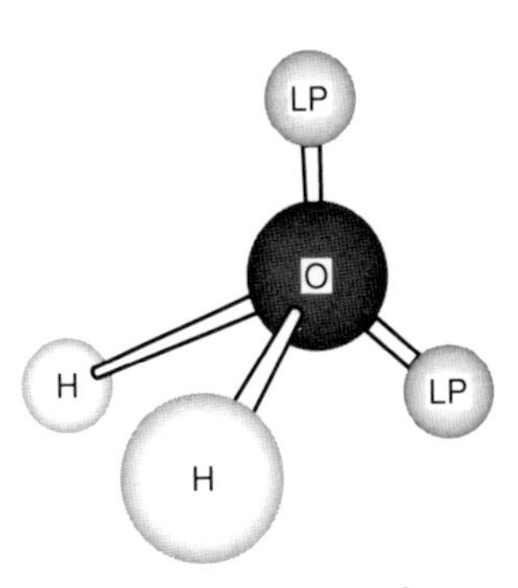

The geometry of water, OH_2

Boron trifluoride, BF_3

Lewis structure: F—B—F with a third F bonded below B

central atom: **boron**	
valence electrons on central atom:	3
3 F each contribute 1 electron:	3
total:	6
divide by 2 to give electron pairs:	3
3 electron pairs: *trigonal* geometry	

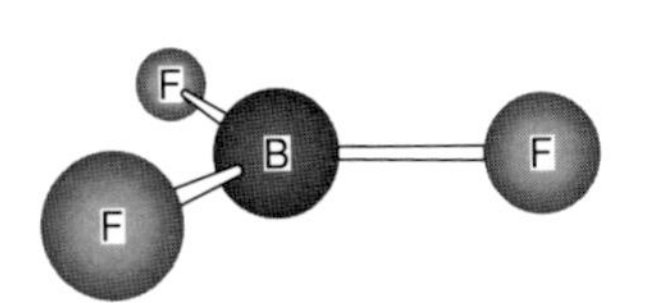

The geometry of boron trifluoride, BF_3

and BF_3 is sometimes written $H_3N{\rightarrow}BF_3$. There are two atoms of interest, N and B. As in any other molecule containing more than one 'central' atom, it is necessary to execute a calculation for *each* centre.

$$\begin{array}{ccccc} & H & & F & \\ H- & N^{+} & - & B^{-} & -F \\ & H & & F & \end{array}$$

An alternative representation of BF_3NH_3 places a positive charge on nitrogen and a negative charge on fluorine. The octet rule is still satisfied for each atom and it now is not necessary to write in an arrow to distinguish the BN bond from other single bonds. In some ways this is an advantage since it removes the impression that a dative bond is different from other bonds. The VSEPR calculation works perfectly well on this representation.

For hexafluorophosphate, $[PF_6]^-$, there are six bonded groups and so no

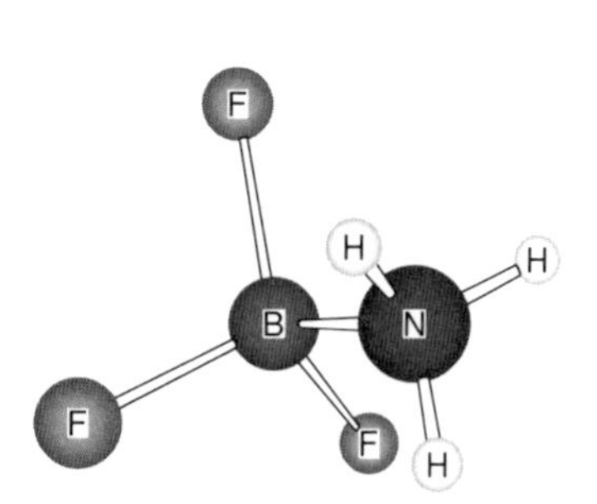

The geometry of $H_3N{\rightarrow}BF_3$

$H_3N{\rightarrow}BF_3$

Lewis structure:

$$\begin{array}{ccccc} & H & & F & \\ H- & N & \rightarrow & B & -F \\ & H & & F & \end{array}$$

central atom: **boron**	
valence electrons on central atom:	3
3 F each contribute 1 electron:	3
NH_3 donates 2 electrons in dative bond:	2
total:	8
divide by 2 to give electron pairs:	4
4 electron pairs: *tetrahedral* geometry at boron	
central atom: **nitrogen**	
valence electrons on central atom:	5
3 H each contribute 1 electron:	3
BF_3 donates 0 electrons in dative bond:	0
total:	8
divide by 2 to give electron pairs:	4
4 electron pairs: *tetrahedral* geometry at nitrogen	

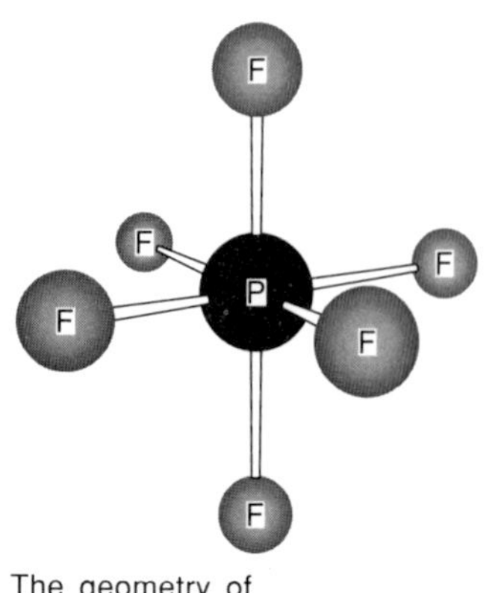

The geometry of hexafluorophosphate, $[PF_6]^-$

Hexafluorophosphate, $[PF_6]^-$

Lewis structure: PF_6^- (P bonded to six F, negative charge on P)

central atom: **phosphorus**	
valence electrons on central atom	5
6 F each contribute 1 electron	6
add one for the negative charge on P	1
total:	12
divide by 2 to give electron pairs:	6
6 electron pairs: *octahedral* geometry	

lone pairs. This anion is useful in synthesis since it often aids the crystallization of bulky cations by providing a reasonable size match for the cation. Note that the negative charge for the purposes of the calculation is placed on phosphorus for the purpose of the calculation even though the negative charge is in reality delocalized over all seven atoms of the ion.

There are four σ-bonded phenyl groups attached to arsenic in the tetraphenylarsonium cation and the calculation predicts no lone pairs on the central atom. The tetraphenylarsonium cation is also useful in synthesis, this time for its ability to aid the crystallization of bulky anions by providing a good size match. Note that the positive charge is placed formally on arsenic, even though it is in reality delocalized over the whole molecule.

For propene the calculation again predicts no lone pairs on the central carbon. Here the VSEPR treatment is seen for a double bond, and note the VSEPR method works for organic species just as well as it does for inorganic molecules. The π electrons in the double bond cause that vertex to appear a little bigger than a simple σ bond and so to make room, the Me—C=CH_2 bond angle (124.8°) is a little greater than the ideal angle of 120°.

Tetraphenylarsonium, $[AsPh_4]^+$

Lewis structure:

```
       Ph
       |
   Ph- As+- Ph
       |
       Ph
```

central atom: **arsenic**	
valence electrons on central atom:	5
4 Ph groups each contribute 1 electron:	4
subtract one for the positive charge on As:	–1
total:	8
divide by 2 to give electron pairs:	4
4 electron pairs: *tetrahedral* geometry	

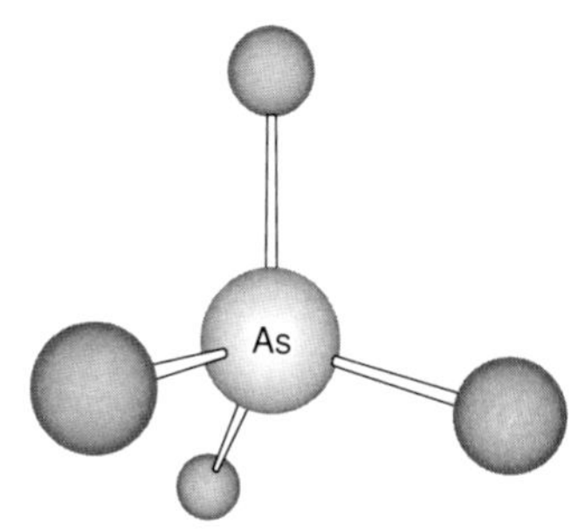

The geometry of the tetraphenylarsonium cation, $[AsPh_4]^+$. Only the attached phenyl carbon atoms are shown.

Propene, $MeCH{=}CH_2$

Lewis structure:

```
        CH2
        ||
   Me — C — H
```

central atom: **carbon** (Me**C**H=CH_2)	
valence electrons on central atom:	4
1 Me groups contributes 1 electron:	1
1 H atom contributes 1 electron:	1
1 =CH_2 group contributes 1 electron in a *s* bond:	1
subtract one for the electron contributed by C to the *p* bond:	–1
total:	6
divide by 2 to give electron pairs:	3
3 electron pairs: *trigonal* geometry	

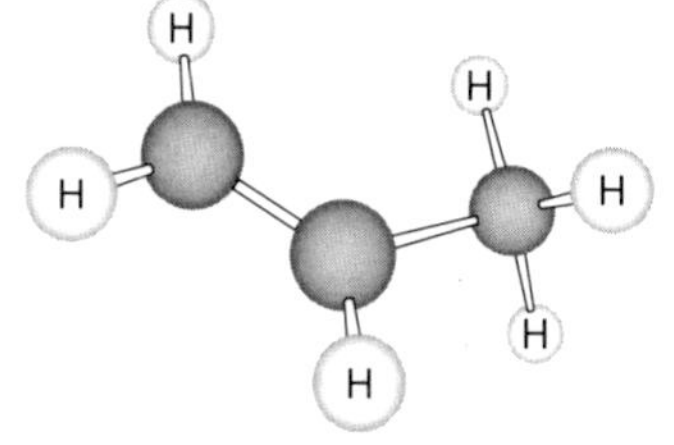

The geometry of propene, $MeCH{=}CH_2$

Note that the net effect of a triple bonded group such as ≡N is –1, made up from +1 in the σ bond and –2 for the two π bonds. As for a double bond, the S≡N bond takes up more room than a single bond and this causes the $F_3S{\equiv}N$ F—S—F bond to drop a little below the ideal tetrahedral bond angle.

The case of ClF_3 is interesting. The calculation shows that the shape is based upon five electron pairs and the favoured geometry is therefore trigonal bipyramidal. There are three bonded groups and so two lone pairs. This is indeed the case, but the point of interest here is the location of the lone pairs. There are three possible ways of placing two electron pairs in a trigonal bipyramidal geometry.

a *b* *c*

These three structures have respectively zero, one, and two lone pairs in the axial sites. For the VSEPR method to be worth much, it has to successfully predict the correct geometry. To approach this problem it is

The geometry of $F_3S{\equiv}N$

Trifluorothionitrile, $F_3S{\equiv}N$

Lewis structure: F—S(≡N)(—F)—F

central atom: **sulphur**	
valence electrons on central atom:	6
3 F atoms each contribute 1 electron:	3
1 terminal nitrogen contributes 1 electron in a *s* bond:	1
subtract *two* for the *two* electrons contributed by S to the *two p* bonds:	–2
total:	8
divide by 2 to give electron pairs:	4
4 electron pairs: *tetrahedral* geometry	

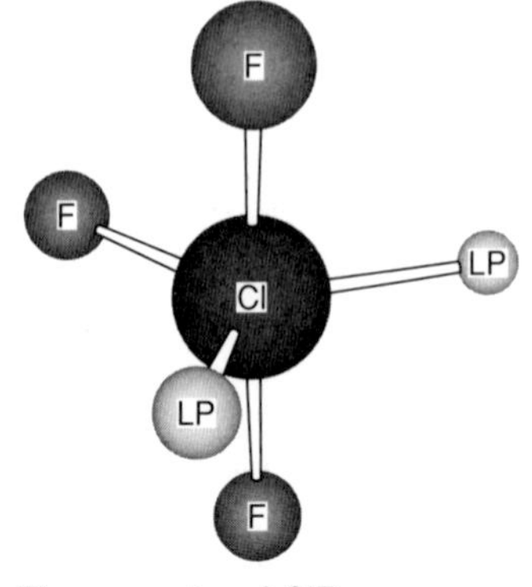

The geometry of ClF_3

Chlorine trifluoride, ClF_3

Lewis structure: F—Cl(—F)—F

central atom: **chlorine**	
valence electrons on central atom:	7
3 F atoms each contribute 1 electron:	3
total:	10
divide by 2 to give electron pairs:	5
5 electron pairs: *trigonal bipyramidal* arrangement	

Table 4.3. Electron pair interactions in ClF_3

Structure	90°	120°	180°
a	2 *bp-bp*	1 *lp–lp*	1 *bp–bp*
	4 *bp-lp*	2 *lp-bp*	
b	2 *bp–bp*	1 *bp–bp*	1 *lp–bp*
	3 *lp–bp*	2 *lp–bp*	
	1 *lp–lp*		
c	6 *lp–bp*	3 *bp–bp*	1 *lp–lp*

necessary to know the relative magnitude of the various kinds of electron pair–electron pair interactions. There are three possible interactions.

lone-pair ↔ lone-pair	(*lp–lp*)
lone-pair ↔ bond-pair	(*lp–bp*)
bond-pair ↔ bond-pair	(*bp–bp*)

Recall that lone pairs occupy more angular space and are located closer to their atoms than bond pairs. Given this, one can argue that the interactions between two lone pairs at any given angle are *greater* than the interactions between two bonding pairs. The interaction between a lone pair and a bonding pair is somewhere in between. Any given structure will adopt a configuration in which interactions between electron pairs are minimized. So, to make a comparison between the three plausible geometries for ClF_3, it is necessary to compare the repulsions between the electron pairs in each of the three cases. A trigonal bipyramid has six interactions at 90°, three at 120° and one at 180°. The interactions for isomers *a*, *b*, and *c* are listed in Table 4.3.

To a first approximation, one can disregard all the interactions that are greater than 90°. The effective distance between electron pairs for 120° and 180° angles is sufficiently large that their mutual interactions are small relative to the 90° interactions. This means that there are only the six interactions to address. Inspect the list of interactions in the 90° column and note that there is a common factor of three *lp–bp* interactions for each. These three interactions can be 'cancelled out', leaving comparison of just three interactions (Table 4.4). Consider structures *a* and *b*. Relate two *bp–bp* + one *bp–lp* interactions for *a* with two *bp–bp* + one *lp–lp* interactions for *b*. The two *bp–bp* interactions 'cancel' to leave a comparison of one *lp–bp* with one *lp–lp* interaction. The *lp–lp* interaction is less stable and it is therefore concluded that *a* is more stable than *b*.

Now consider structures *a* and *c*. This involves comparing one *lp–bp* + two *bp–bp* interactions with three 3 *lp–bp* interactions. One *lp–bp* interaction is 'cancelled' out for each leaving a comparison of two *bp–bp* with two *lp–bp* interactions. Since *bp–bp* interactions are less unfavourable than *lp–bp* interactions, structure *a* is more favourable than *c*.

Since *a* is more favoured than either *b* or *c*, structure *a* is the most favoured structure. The geometry of ClF_3 is therefore based upon a trigonal bipyramidal arrangement of five electron pairs but the 'visible shape' is 'T-shaped', as is indeed the case. Both electron pairs are equatorial.

Table 4.4. After 'cancellation' and removal of interactions > 90°

structure	90°
a	2 *bp–bp*
	1 *bp–lp*
b	2 *bp–bp*
	1 *lp–lp*
c	3 *lp–bp*

Since the lone pairs occupy a little more than their fair share of the available space, the effect is for the three Cl—F bonds to squeeze together slightly to make room for the lone pairs. The trigonal bipyramid is therefore slightly distorted with the Cl_a—F—Cl_e bond angle less (87.5°) than 90°.

The perchlorate example illustrates the fact that the correct geometry is predicted even when terminal oxygen is assumed formally to bond always as =O, giving a total of 'eight' bonds to Cl. This is nowhere clearer than in the calculation for the carboxyl carbon in ethanoate (acetate) for which the modified Lewis structure includes a '5-valent' carbon (procedure 3, page 59), but nevertheless generates the correct answer.

Another complication crops up when there are *unpaired* electrons. This is well illustrated for nitrogen dioxide. So, for NO_2 there is an integral number of electrons but a non-integral number of electron pairs. Since any orbital can

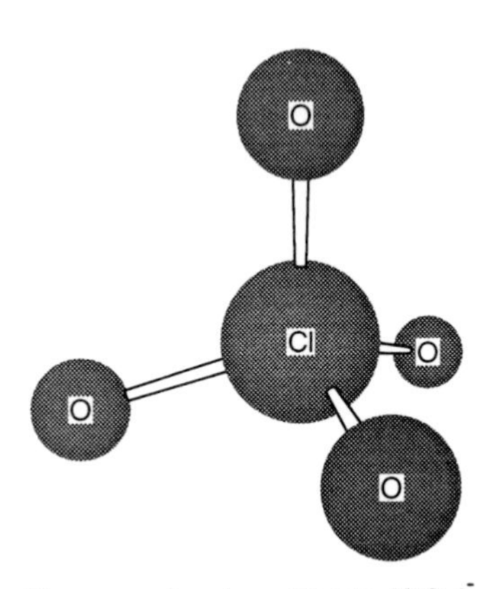

The geometry of perchlorate, $[ClO_4]^-$

Perchlorate, $[ClO_4]^-$

Lewis structure: O=Cl(=O)(=O)=O (with negative charge on Cl)

central atom: **chlorine**	
valence electrons on central atom:	7
4 terminal oxygens each contribute 1 electron in the four σ bonds:	4
subtract four for the four electrons contributed by Cl to the four π bonds (one for each):	–4
add one for the negative charge located on Cl:	1
total:	8
divide by 2 to give electron pairs:	4
4 electron pairs: *tetrahedral* geometry	

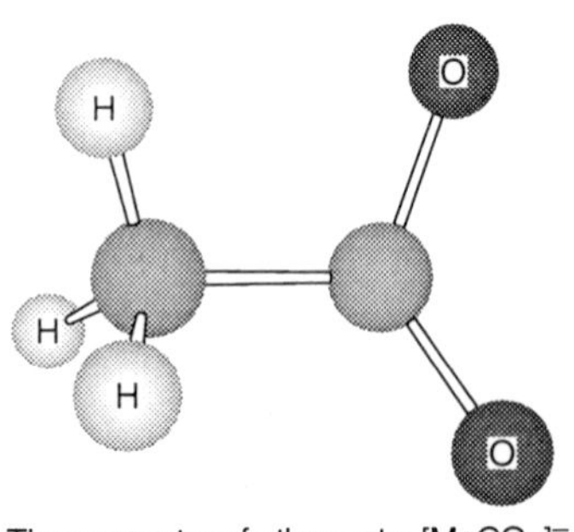

The geometry of ethanoate, $[MeCO_2]^-$

Ethanoate (acetate), $[MeCO_2]^-$

Lewis structure: Me—C(=O)=O (with negative charge on C)

central atom: **carbon**	
valence electrons on central atom:	4
1 Me group contributes 1 electron:	1
2 terminal oxygens each contribute 1 electron in the two σ bonds:	2
subtract two for the two electrons contributed by C to the two π bonds:	–2
add one for the negative charge located on C:	1
total:	6
divide by 2 to give electron pairs:	3
3 electron pairs: *trigonal* geometry	

Nitrogen dioxide, NO_2	
Lewis structure: O=N=O	
central atom: **nitrogen**	
valence electrons on central atom:	5
2 terminal oxygens each contribute 1 electron in the two σ bonds:	2
subtract two for the two electrons contributed by N to the two π bonds:	–2
total:	5
divide by 2 to give electron pairs:	$2^1/_2$
$2^1/_2$ electron pairs: *trigonal planar* arrangement	

The geometry of nitrogen dioxide, NO_2

accommodate 0, 1, or 2 electrons, $2^1/_2$ electron pairs must be placed into *three* orbitals. Therefore the geometry is based upon a trigonal-planar arrangement of electron pairs. Since the lone-pair orbital is only half filled, it demands less space, and the O—N—O angle opens out a little (to 134.1°) from the ideal trigonal angle of 120°.

Addition of one electron to NO_2 gives the nitrite anion NO_2^-. This last electron completes the half-occupied lone-pair orbital and this filling of the orbital causes it to fill out, and so close the O—N—O bond angle to 115°.

4.2 VSEPR and more than six electron pairs

It is considerably less easy to draw a distinction between apparently reasonable seven coordinate geometries. There are several possibilities, including the pentagonal bipyramid and the capped octahedron (Fig. 4.2).

Iodine heptafluoride, IF_7, is a good example of a pentagonal bipyramidal geometry. The molecule XeF_6 is an interesting case. As with IF_7, application of VSEPR rules suggests seven electron pairs. These are made up from six bonding pairs and one lone pair. In fact, the structure of XeF_6 is based upon a distorted octahedron, probably towards a monocapped octahedron. It is difficult to settle the geometry of the lowest energy configurtion because the geometry of XeF_6 changes rapidly with time, that is, it is *fluxional*. The effect of this fluxional process is to average all the

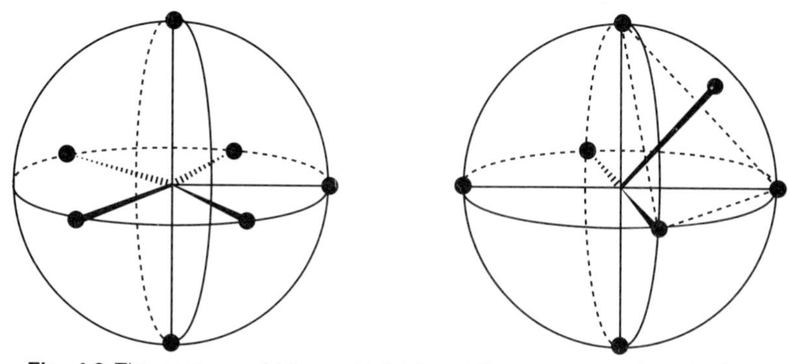

Fig. 4.2 The pentagonal bipyramid (left) and the monocapped octahedron.

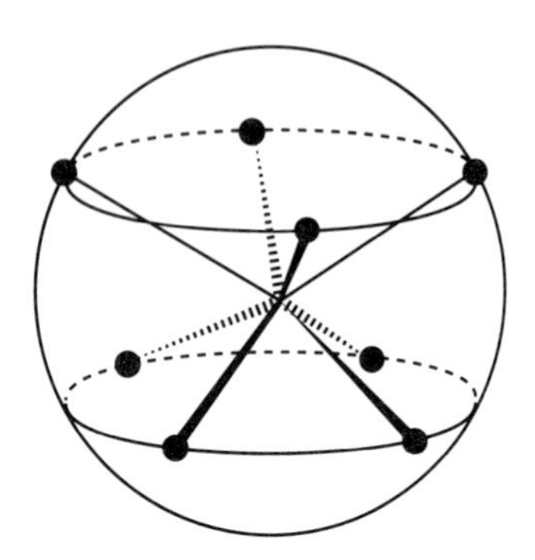

Fig. 4.3 The idealized coordination geometry for eight electron pairs

fluorine positions.

With higher coordination numbers the situation is more complex. For instance, the idealized geometry for eight electron pairs (epitomized by the anion $[XeF_8]^{2-}$) is a square antiprism (Fig. 4.3) but the energy of other coordination geometries may be very similar or more stable in particular cases.

4.3 Transition metal systems

The method delineated above works very well for main group compounds but is less useful for transition metal compounds. The problem is that the lone pairs in transition metal complexes do not occupy directional orbitals in the same way as those in main group compounds. One can count electrons for transition metal compounds in exactly the same way as for main group compounds but typically the number of valence electrons will be rather larger than observed for main group metal compounds. Typical observed numbers of valence electrons vary with actual geometry. For instance, one might find anything from 12 to 22 valence electrons as the electron count in octahedral complexes. The first twelve of these are used for bonding to the six ligands. The 13th and subsequent electrons up to the 22nd, however do not, to a first approximation at least, affect the shape of the complex. They lie in atomic *d* orbitals which are not strongly shape-determining.

In fact, despite the larger number of electrons involved, the first order prediction of shape for transition metal complexes is *simpler* than for main group compounds. One can obtain very good results by simply counting *ligands* and placing the attached ligand groups as far apart from each other as possible. Six ligands therefore adopt an octahedral configuration, five a trigonal-bipyramidal configuration, and four a tetrahedral configuration.

There are exceptions however. There is an important group of 16-electron compounds with four attached ligands and which are d^8 metals. Many of these are square-planar complexes. There is also an interesting group of complexes which are d^0 ML_6 species. These tend to be trigonal-prismatic rather than octahedral.

4.4 Exercises

1. Use VSEPR rules to predict the shapes and geometries of the central atoms of the following: BeH_2, $[BeF_4]^{2-}$, SF_6, $[IO_3]^-$, $[IO_4]^-$, $[IO_6]^{5-}$, $[GaBr_4]^-$, $[NCS]^-$, $[SnPh_3]^-$, SO_2, SO_3, $SOCl_2$, SO_2Cl_2, MeCN, HCCH, H_2CCH_2, CO_2, $[CO_3]^{2-}$, $[NO_3]^-$, NH_2^-, $XeOF_4$, $[SbCl_6]^-$, $[H_3O]^+$, $[NH_4]^+$, IF_5, $[AlCl_6]^{3-}$, Me^-, $SbPh_5$, $[SO_4]^{2-}$, O_3, Me_2SO, PCl_3, and $[SbO_4]^{3-}$. Confirm for which compounds your predicted geometries are correct by referring to appropriate textbooks.
2. Use the VSEPR rules to predict the coordination geometries of XeF_2, $[IF_4]^-$, and SF_4. Write down reasonable isomers for your predicted geometries and determine in each case which isomer is the most likely.

5 Hybrid orbital description of bonding

The use of Lewis structures and the VSEPR method allow prediction of the shape of a molecule, together with the number of lone pairs and bond pairs. What these methods do not explain is *how* the electrons are involved in the lone and bonding pairs. There are two principal ways of proceeding. One involves regarding bonds as *localized interactions involving two electrons between two atoms* while the other involves *delocalization of electrons in molecular orbitals delocalized over the whole molecule.*

The *localized* bond approach is called the *valence bond approach.* The *delocalized* bond approach is executed in the *molecular orbital approach.* The valence bond approach was developed first and involves the conceptual bringing together of atoms, which are then allowed to interact. The molecular orbital approach involves placing the nuclei and core electrons into position, and then placing all the valence electrons into multicentre molecular orbitals. Neither method is exact. Often, the molecular orbital approach is more appropriate for spectroscopic discussions. Using more sophisticated mathematical approaches to the description of bonding one can make both the valence bond and molecular orbital approach more and more closely the 'exact' wave function, as measured by correspondence of calculated energy to experimentally determined energy. In the limit, therefore, both descriptions are equivalent.

A localized bond approach implies that one orbital on one atom interacts with an orbital on a second atom to form a bond. Orbitals on other atoms in the molecule are not involved. This works for hydrogen as follows. To a first approximation, the H—H bond is viewed as an overlap between the hydrogen 1*s* orbitals on each atom. In practice this is only a first approximation. There are many alterations which can be made to the 1*s* wave function to get a more accurate wave function for the molecule, but in particular, an improvement is found once it is recognized that the second hydrogen nucleus (positively charged) causes a distortion of the electron density in the first hydrogen atom's 1*s* orbital. Mathematically, this is conveniently described by mixing and normalizing the mathematical function describing the 1*s* orbital with a *small* component (only about 1%) of the mathematical description of the hydrogen $2p_z$ orbital. This process is called *hybridization.* The effect on the pictorial description of the hydrogen 1*s* orbital is to make an orbital which is distorted slightly towards the nucleus of the second hydrogen atom. This improves the extent of overlap with the correspondingly distorted orbital on the second atom and the calculated bond energy is a better approximation to the experimentally determined bond energy.

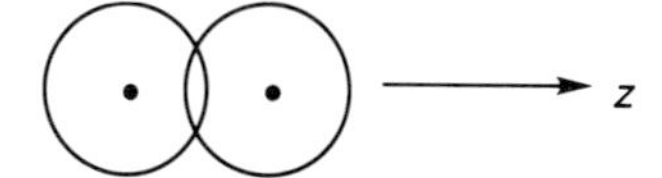

Fig. 5.1 The overlap of two hydrogen 1*s* orbitals.

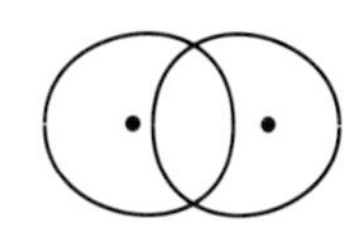

Fig. 5.2 The overlap of two hydrogen 1*s* orbitals after mixing with about 1% of a $2p_z$ function.

5.1 Hybridization: linear systems

Linus Pauling: 1901–

Hybridization, or mixing of orbitals on an atom, is a useful mathematical device. It is not something that atoms *do* in the sense that hybridization does not occur during chemical reactions, it is a convenient mathematical procedure which describes bonding in terms of conceptually simple two-centre, two-electron bonds. It also gives a way of forming atomic orbitals which overlap more strongly with atomic orbitals on other atoms to form more stable bonding molecular orbitals. It was developed by Pauling. The concept of hybridization is best discussed with reference to a few examples.

Beryllium dihydride

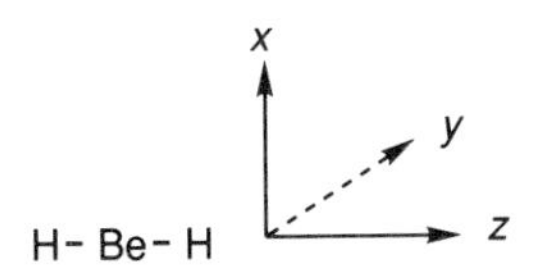

Prediction of the geometry of BeH_2 by VSEPR rules correctly suggests that BeH_2 is linear. By convention, in a linear molecule, the axis of the molecule is set to the z-axis. The ground state configuration of beryllium is $[He]2s^2$. The problem is to describe two bonds arranged at 180° to each other. There are no unpaired electrons on the beryllium atom in the ground state with which to form two shared electron-pair bonds. To construct these bonds, it is necessary to start with *two* orbitals on beryllium, *each* containing one electron. Bond formation then proceeds by pairing up the electrons with the two electrons from the two H atoms. Two unpaired electrons are achieved by 'promoting' one of the beryllium electrons into a $2p_z$ orbital (Eqn 5.1). This is a hypothetical excitation, which while convenient for this explanation, is not a process which actually happens on formation of BeH_2.

$$[He]2s^2 \xrightarrow{\text{promotion}} [He]2s^1 2p_z^1 \tag{5.1}$$

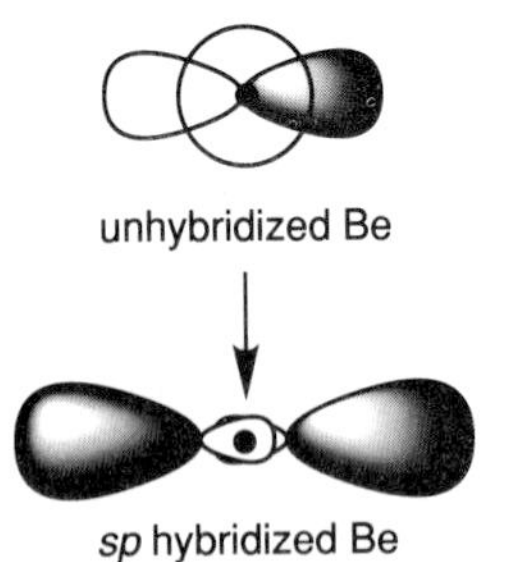

Fig. 5.3 Formation of two *sp* hybrid orbitals from the 2*s* and $2p_z$ Be orbitals.

The problem now is that while the $2p_z$ orbital points in the correct direction (the molecule's axis), the 2*s* orbital is non-directional. This would suggest one bond along the z-axis with another bond at some undefined direction. The way forward is to do a mathematical construction in which the 2*s* orbital is 'mixed' with the $2p_z$ orbital. The result of this *hybridization* process is a set of two orbitals which point in the correct directions (Fig. 5.3). The two resulting orbitals are equivalent and each of the hybrids contain 50% *s* character and 50% *p* character. In this case, because the hybridization involves mixing one s and one p orbital, the hybrid orbitals are called *sp* hybrids. The hybridization process is represented by Eqn 5.2.

$$[He]2s^1 2p_z^1 \xrightarrow{\text{hybridization}} [He](sp)^2 \tag{5.2}$$

$$sp(1) = \frac{1}{\sqrt{2}}(2s + 2p_z) \qquad sp(2) = \frac{1}{\sqrt{2}}(2s - 2p_z) \tag{5.3}$$

The principal quantum numbers are not usually written in the resulting hybrids, it is usually clear which orbitals are involved from the context of the molecule under discussion. The symbol $(sp)^2$ refers to two *sp* hybrids containing a total of two electrons, one in each. The *sp* hybrid orbital has cylindrical symmetry meaning that it can be referred to as a σ orbital. Each of the two hybrid orbitals is populated by one electron. Mathematically, each is generated by taking linear combinations of the 2*s* and $2p_z$ orbitals (Eqn 5.3).

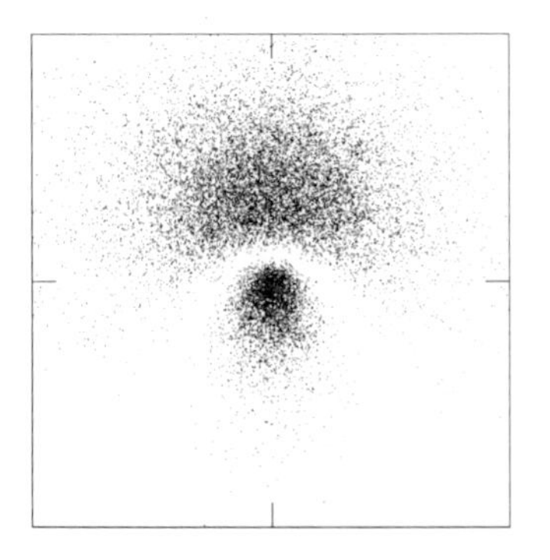

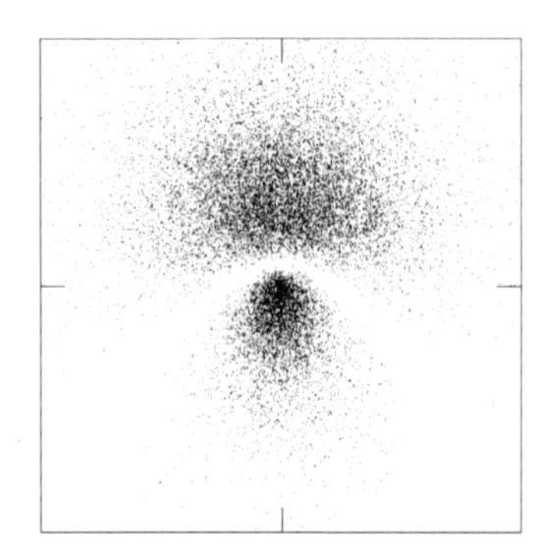

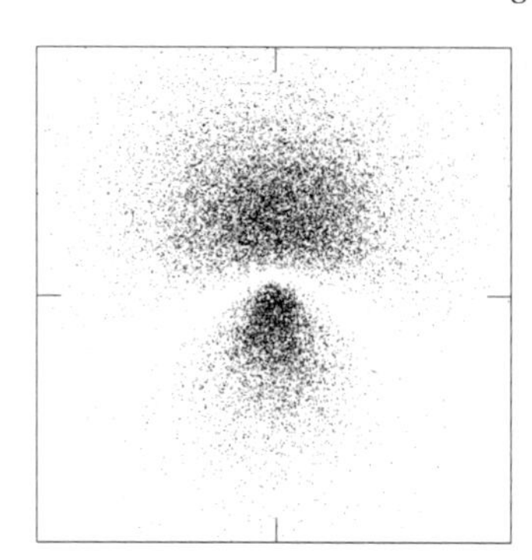

Fig. 5.4 Cross-section electron dot density diagrams (15 000 dots) representing an *sp* hybrid orbital (left), an sp^2 hybrid orbital (centre), and an sp^3 hybrid orbital (right) for carbon. The nucleus lies at the centre of the squares.

The electron dot density representation of one of the *sp* hybrids (Fig. 5.4, left) shows a major lobe directed in one direction with a lesser lobe, where the wave function has the opposite sign, in the other direction. There is a curved nodal surface.

Note the position of the beryllium nucleus. The nucleus resides *inside* the minor lobe and *not* on the nodal surface as frequently represented in books. In the schematic representations of hybrid orbitals (Fig. 5.5), the nucleus is correctly placed in the right-hand diagram, but not so in the left-hand diagram.

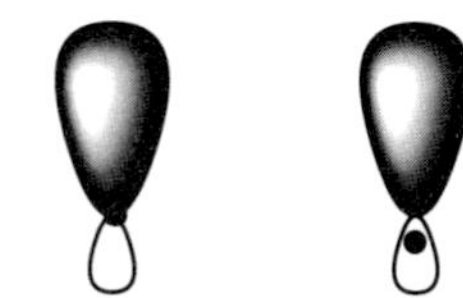

Fig. 5.5 In these two representations of hybrid orbitals, the nucleus is incorrectly placed in the left-hand diagram

The two H atoms bond to the Be atom by forming two two-electron bonds through overlap of each *sp* hybrid, each containing one electron, with the hydrogen 1*s* orbitals, each of which contains one electron.

2H• + B̈e ⟶ H:Be:H

Under the valence orbital approach, the bonding in beryllium dihydride therefore consists of two shared electron bonds. Since each Be—H bond is a separate entity under the hybridized orbital approach, the analogy with the Lewis representation is clear.

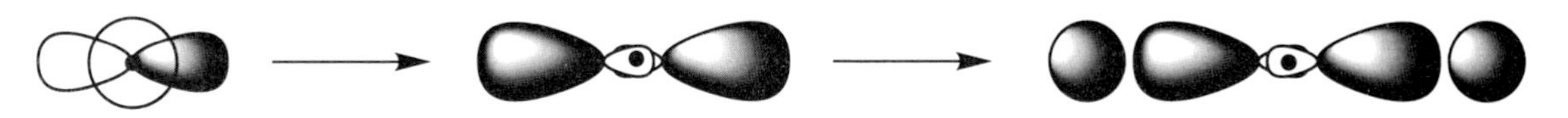

Unhybridized Be | *sp* hybridized Be | Orbital overlaps to form BeH_2

Fig. 5.6 Beryllium *sp* hybridization and presentation to two H 1*s* orbitals to form BeH_2.

Ethyne

Lewis structure: H:C:::C:H

Ethyne (acetylene) is correctly predicted to be linear by the VSEPR method. The connectivity is H—C—C—H. The Lewis structure of ethyne shows a triple bond between the two carbon atoms. Therefore, a description of the bonding in which each carbon forms one shared electron-pair bond with one hydrogen atom and three shared electron bonds to the carbon atom is required. The electronic configuration of carbon is $[He]2s^22p^2$. The two 2*p* electrons reside in separate orbitals so two *p* orbitals have one electron each, one *p* orbital has none and the *s* orbital has two. The first move is to conceptually promote an electron from the 2*s* level to the vacant 2*p* level (Eqn 5.4). The next stage is hybridization (Eqn 5.5). As for BeH_2, the 2*s* orbital is mixed

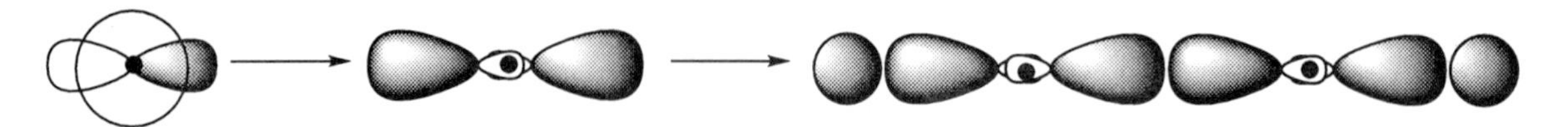

Fig. 5.7 The σ framework in ethyne

with the $2p_z$ orbital. This leaves the $2p_x$ and $2p_y$ orbitals unhybridized and still possessing one electron each. (5.4)

$$[\mathrm{He}]2s^2 2p^2 \xrightarrow{\text{promotion}} [\mathrm{He}]2s^1 2p_x^1 2p_y^1 2p_z^1$$

$$[\mathrm{He}]2s^1 2p_x^1 2p_y^1 2p_z^1 \xrightarrow{\text{hybridization}} [\mathrm{He}](sp)^2 2p_x^1 2p_y^1 \qquad (5.5)$$

One of the resulting *sp* hybrids is used to bond to the hydrogen atom (Fig. 5.7). The second is used to overlap with an equivalent *sp* hybrid orbital on the second C atom to make a C—C bond. Both these are again σ bonds.

There are two *p* orbitals remaining, each containing a single electron. These p_x and p_y orbitals are used to construct two orbitals with π symmetry (Fig. 5.8). These are two equivalent orbitals at 90° to each other. Note the relationship between these orbitals and the π orbitals of N_2 (Section 3.13). As presented here, it looks as though ethyne should have four lobes directed out along the *x*- and *y*-axes when viewed down the HC≡CH axis. In fact the mathematics of the situation is such that the symmetry of the two π bonds when viewed *together* is completely *cylindrical* (Fig. 5.9).

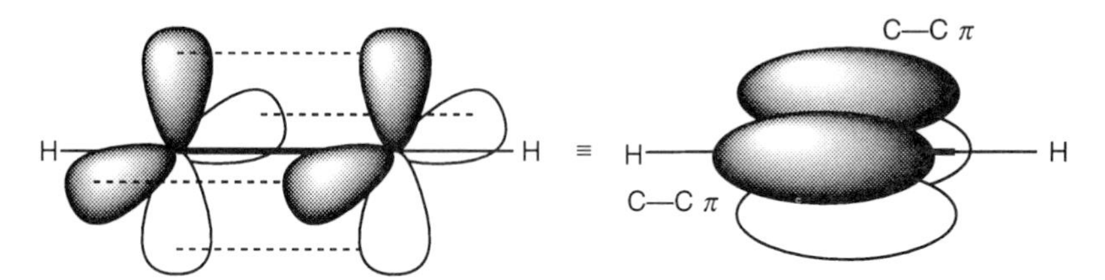

Fig. 5.8 The two π bonds in ethyne.

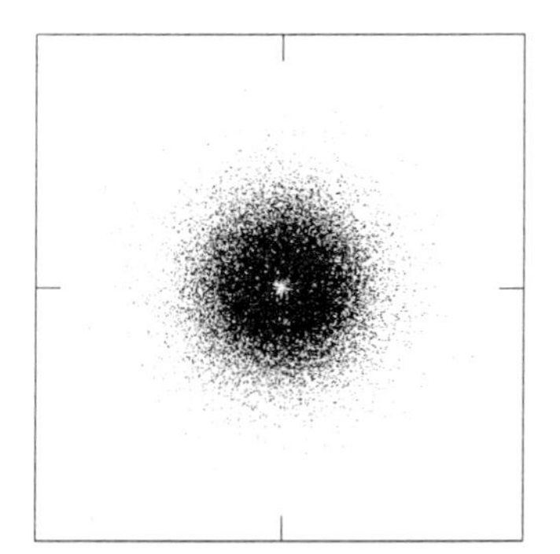
Fig. 5.9 Electron dot density plot for two π orbitals viewed along the HC≡CH axis.

5.2 Hybridization: trigonal systems

Atoms within molecules with a trigonal geometry are those for which prediction by the VSEPR method predicts their shape to be determined by three electron pairs. The most convenient way of generating acceptable trigonal hybrid orbitals is to hybridize one *s* with two *p* orbitals. These orbitals are called sp^2 hybrids. The three resulting orbitals are equivalent and each of the hybrids contain 33.3% *s* character and 66.7% *p* character. One carbon sp^2 hybrid is illustrated in Fig. 5.4 (centre).

Boron trihydride

The geometry of boron in BH_3 is correctly predicted by the VSEPR method as trigonal. In this case, by convention the z-axis is taken to be perpendicular to the plane of the molecule. The electronic configuration of boron is $[He]2s^22p^1$. Promotion of one of the $2s$ electrons to a $2p$ orbital (Eqn 5.6) gives three valence orbitals, each of which is populated by one electron.

The axis definition for BH_3.

$$[He]2s^2 2p^1 \xrightarrow{\text{promotion}} [He]2s^1 2p_x^1 2p_y^1 \quad (5.6)$$

$$[He]2s^1 2p_x^1 2p_y^1 \xrightarrow{\text{hybridization}} [He](sp^2)^3 \quad (5.7)$$

It makes sense to use the $2p_x$ and $2p_y$ orbitals since the $2p_z$ orbital is directed out of the plane of the three bonds. Of the three orbitals containing one electron, two are at 90° to each other and the other is non-directional. This is not convenient for constructing three trigonal bonds (120° bond angles). Mathematically, the way forward is to hybridize these three orbitals into trigonal hybrids (Eqn 5.7). These are called sp^2 hybrids since they arise from one s and two p orbitals. The remaining $2p$ orbital ($2p_z$) is both empty and non-bonding.

The sp^2 hybrid orbitals are now set up to overlap with the three hydrogen atoms. Each bond is a σ bond. Again there is a correspondence of this structure with the Lewis structure.

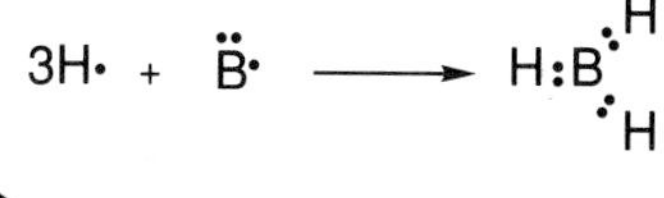

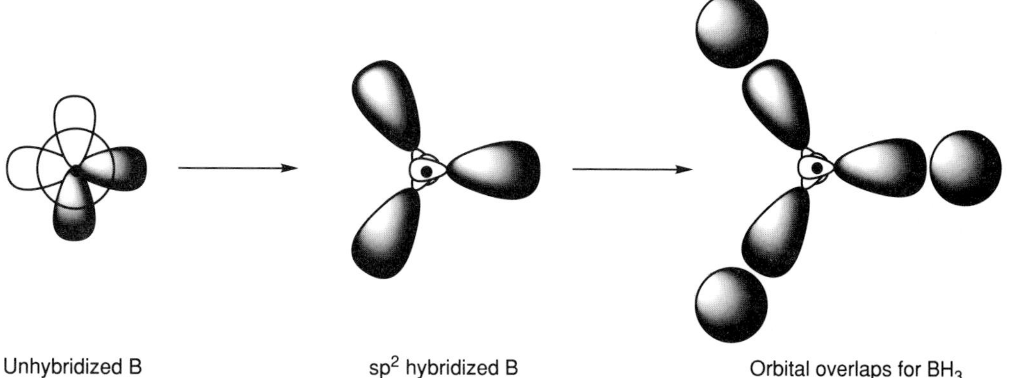

Fig. 5.10 The overlap of boron sp^2 hybrids with hydrogen $1s$ orbitals to form BH_3.

Ethene

The VSEPR prediction is that the carbon atoms in ethene are approximately trigonal (a planar geometry with bond angles around 120°). The z-axis is defined here along the CC axis with the x-axis perpendicular to the plane of the molecule. Promotion of an electron from the $2s$ orbital into an empty p orbital gives four carbon orbitals, each containing one electron (Eqn 5.8). To construct trigonal hybrids, the approach is then to mix the $2s$ orbital with two of the p orbitals, the p_y and p_z orbitals with the axis definition used here (Eqn 5.8).

$$[He]2s^2 2p^2 \xrightarrow{\text{promotion}} [He]2s^1 2p_x^1 2p_y^1 2p_z^1 \quad (5.8)$$

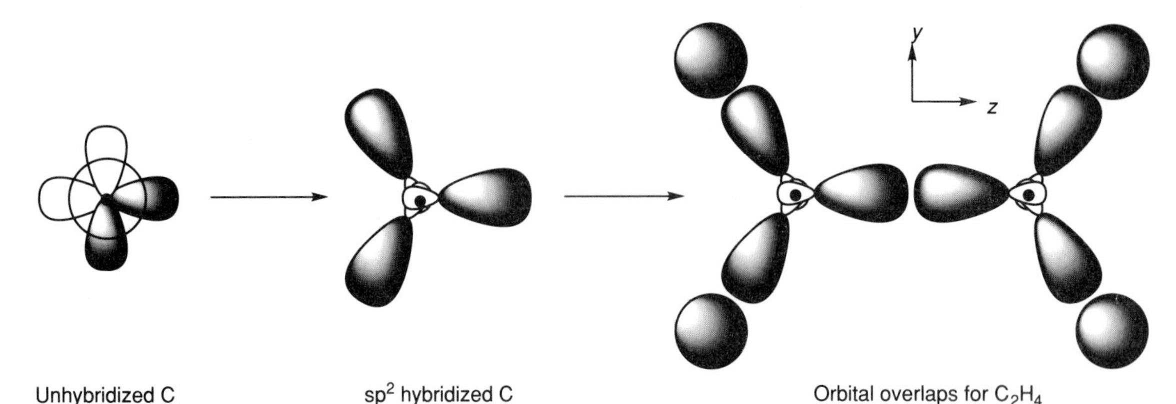

Fig. 5.11 The σ framework in ethene.

This gives three equivalent sp^2 hybrid orbitals. Two of the sp^2 hybrids are used to bond to hydrogen atoms. The third is used to construct a C–C bond by bonding to an equivalent sp^2 hybrid on the second atom. This leaves the remaining $2p_x$ orbital. This $2p_x$ orbital is orientated in such a way that it is set up for making a π bond. Again the valence orbital result generated parallels the Lewis structure.

$$[\mathrm{He}]2s^1 2p_x^1 2p_y^1 2p_z^1 \xrightarrow{\text{hybridization}} [He](sp^2)^3 2p_x^1 \tag{5.9}$$

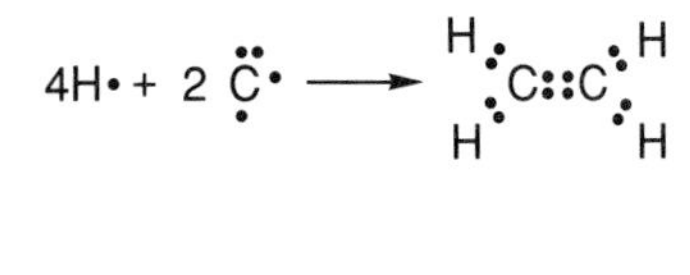

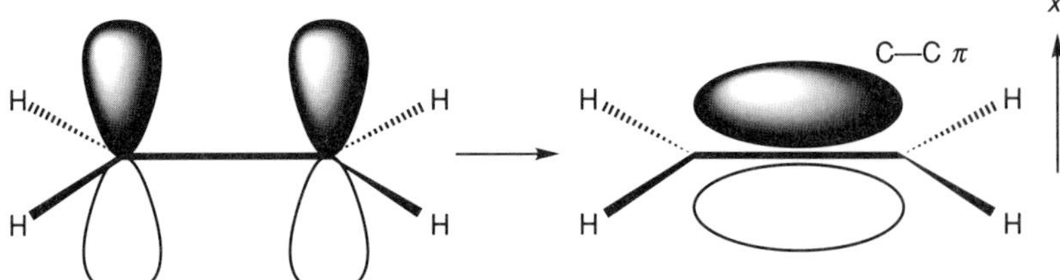

Fig. 5.12 The π bond in ethene.

5.3 Hybridization: tetrahedral systems

Systems based upon a tetrahedral geometry require that hybridization involves four orbitals. The most obvious way to do this involves mixing the *s* and the three *p* valence orbitals. Such hybrids are referred to as sp^3 hybrids. The four resulting orbitals are equivalent and each of the hybrids contain 25% *s* character and 75% *p* character. A carbon sp^3 hybrid orbital is illustrated in Fig. 5.4 (right).

Methane

The ground state electronic structure of carbon is $[\mathrm{He}]2s^22p^2$. Four unpaired electrons are achieved by conceptually promoting one of the electrons into the remaining empty *p* orbital (Eqn 5.10). Hybridization of the four orbitals gives sp^3 hybrids (Eqn 5.11). The mathematics works out so that the orbitals are mutually orientated at angles of 109.5°.

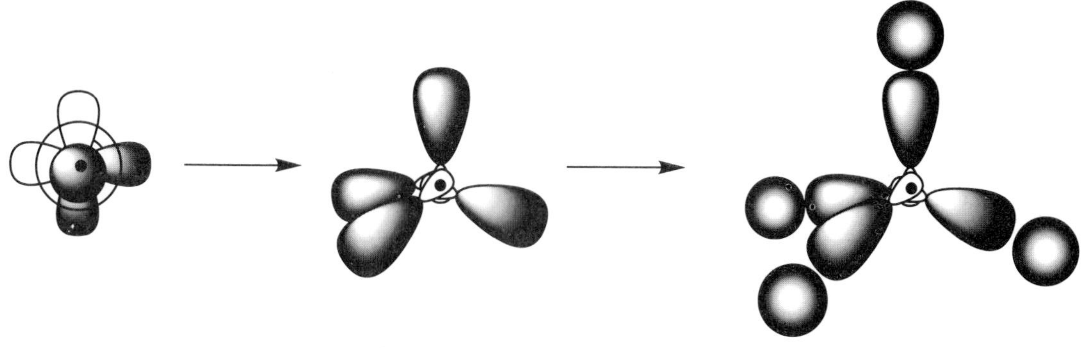

Fig. 5.13 The σ framework in methane.

Note that of the original four carbon orbitals, three are orientated at 90° and the fourth is non-directional. Without invoking hybridization it would be difficult to construct four tetrahedral localized bonds. So in the hybrid view of bonding for methane there are four equivalent C—H bonds with bond angles of 109.5° (Fig. 5.13).

$$[He]2s^2 2p^2 \xrightarrow{\text{promotion}} [He]2s^1 2p_x^1 2p_y^1 2p_z^1 \qquad (5.10)$$

$$[He]2s^1 2p_x^1 2p_y^1 2p_z^1 \xrightarrow{\text{hybridization}} [He](sp^3)^4 \qquad (5.11)$$

5.4 Hybridization: trigonal-bipyramidal systems

Phosphorus pentahydride, PH_5, is a hypothetical molecule, but it serves to illustrate the principle. It is a conceptually simpler system to examine than a real case such as PCl_5. The geometry is predicted to be trigonal bipyramidal by the VSEPR approach. The three *equatorial* hydrogens are orientated at 120° to each other while the two *axial* hydrogen atoms are orientated at 180° to each other and at 90° to the equatorial hydrogen atoms. By convention the z-axis is defined as along the axial bonds. The x- and y-axes lie in the equatorial plane with only one of them able to lie along a bond since it is not possible to map two axes which lie at 90° to each other onto a set of trigonal bonds. Since there are five atoms bonded to the phosphorus, five orbitals on the phosphorus atom are required to form PH_5. The s and the p orbitals only total four. Therefore one of the empty valence d orbitals is used to provide a fifth orbital. The $3d$ orbitals are unoccupied in the ground state of phosphorus. An electron is therefore promoted from the $3s$ orbital to a $3d$ orbital (Eqn 5.12). The $3d_{z^2}$ orbital is most appropriate. The next step is to hybridize mathematically the $3s$, the three $3p$, and the $3d_{z^2}$ orbitals to form five dsp^3 hybrids (Eqn 5.13).

Mathematically, the five hybrid orbitals are arranged in a trigonal-bipyramidal fashion. Since axial and equatorial environments in a trigonal-bipyramidal system are not equivalent, neither are the hybrid orbitals precisely equivalent in this case. This is therefore a slightly different

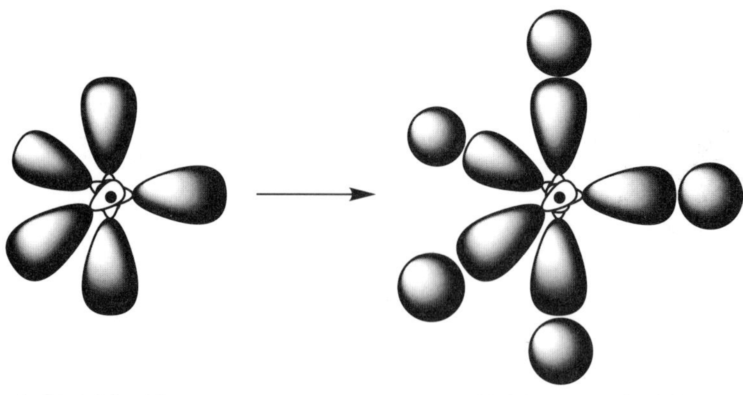

Fig. 5.14 The σ bond framework for PH_5 based upon dsp^3 hybridization of phosphorus.

situation to the hybridization systems discussed above. Each is populated by a single electron. Each H atom 1*s* orbital now combines with one of the hybrid orbitals to form two-electron localized bonds (Fig. 5.14).

$$[\mathrm{Ne}]3s^2 3p^3 \xrightarrow{\text{promotion}} [\mathrm{Ne}]3s^1 3p_x^1 3p_y^1 3p_z^1 3d_{z^2}^1 \quad (5.12)$$

$$[\mathrm{Ne}]3s^1 3p_x^1 3p_y^1 3p_z^1 3d_{z^2}^1 \xrightarrow{\text{hybridization}} [\mathrm{Ne}](dsp^3)^5 \quad (5.13)$$

As before, the hydrogen atoms are now presented to the prepared hybrids to make five σ bonds. It is straightforward to adapt these diagrams to a molecule such as PCl_5. The hybridization of phosphorus is the same as in PH_5, it just remains to decide the nature of the chlorine orbitals that are to be used for bonding to the dsp^3 hybrids. Since the electronic configuration of chlorine is $[Ne]2s^22p^5$, a half-filled *p* orbital directed *towards* the phosphorus would suffice. But an alternative to this is to *sp* hybridize the chlorine to give one hybrid orbital directed *towards* the phosphorus and one orbital directed *away* from phosphorus. The latter orbital contains a lone pair of electrons and the former one electron, and the former is therefore suitable for bonding to the phosphorus. One advantage of this is that it obviates the need to keep an electron pair in a spherical *s* orbital. Instead the lone pair becomes directional when in an *sp* hybrid which perhaps corresponds better with our concept of directional lone pairs. Since the axial and equatorial connected atoms are in different environments, techniques such as NMR spectroscopy show separate signals for axial and equatorial sites under appropriate conditions.

5.5 Hybridization: octahedral systems

Sulphur hexahydride, SH_6 is again a hypothetical molecule, but it also serves to illustrate the principle. As for PH_5 above, SH_6 is simpler to examine than a real molecule such as SF_6. The hexahydride is predicted to be octahedral by the VSEPR method. The six S—H bonds are orientated at 90° to each other and unlike the trigonal-bipyramidal environment, are now all equivalent. Since there are six atoms bonded to the sulphur, six sulphur orbitals are

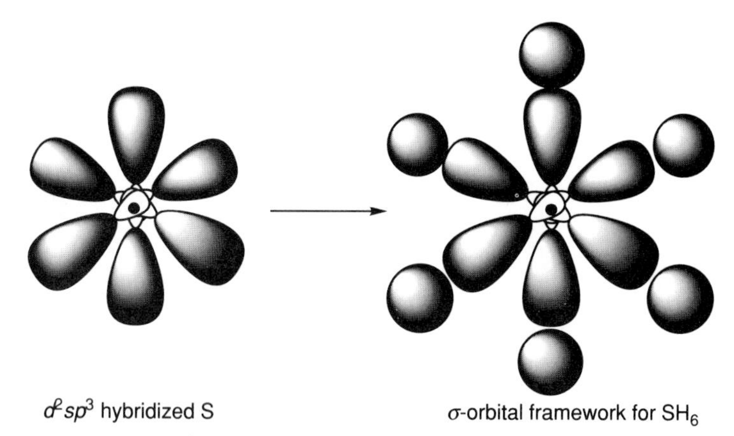

Fig. 5.15 The σ bond framework of SH_6 based upon d^2sp^3 hybridization of sulphur.

required. In this case an electron is promoted from the $3s$ orbital *and* an electron from a p orbital to two of the $3d$ orbitals. Since the $3d_{z^2}$ orbitals and the $3d_{x^2-y^2}$ orbitals already point along the x-, y-, and z-axes, it makes sense to promote into them (Eqn 5.14). The next step is to hybridize mathematically the $3s$, the three $3p$, and the two $3d$ orbitals to form d^2sp^3 hybrids (Eqn 5.15). This hybridization process gives six *equivalent* orbitals arranged octahedrally. Each has a single electron. Each H atom $1s$ orbital now combines with one of the hybrid orbitals to form two-electron localized bonds, resulting in an octahedral structure (Fig. 5.15).

$$[\mathrm{Ne}]3s^2 3p^4 \xrightarrow{\text{promotion}} [\mathrm{Ne}]3s^1 3p_x^1 3p_y^1 3p_z^1 3d_{z^2}^1 3d_{x^2-y^2}^1 \qquad (5.14)$$

$$[\mathrm{Ne}]3s^1 3p_x^1 3p_y^1 3p_z^1 3d_{z^2}^1 3d_{x^2-y^2}^1 \xrightarrow{\text{hybridization}} [\mathrm{Ne}](d^2sp^3)^6 \qquad (5.15)$$

5.6 Benzene

Benzene, C_6H_6, is one of the most important molecules in organic chemistry. It is planar and consists of a hexagonal ring of six carbon atoms, each of which is bonded to one hydrogen atom and two adjacent carbon atoms. The geometry about each carbon is trigonal and so it is convenient to treat each carbon atom as sp^2 hybridized.

A σ bond framework is built up from six sp^2 hybridized carbon atoms (Fig. 5.16), with two of the sp^2 hybrids from each carbon atom used to bond to neighbouring atoms. The third hybrid orbital is then set to bond to a hydrogen atom using the H $1s$ orbital.

However it is the out-of-plane bonding that is of the most interest. The six unhybridized $2p_z$ orbitals are orientated such that they can overlap to form π bonds of the same type displayed by ethene. In a localized model these can form three such π bonds (Fig. 5.17). Therefore a simple localized view of benzene requires that benzene be regarded as cyclohexatriene. This is not sufficient however.

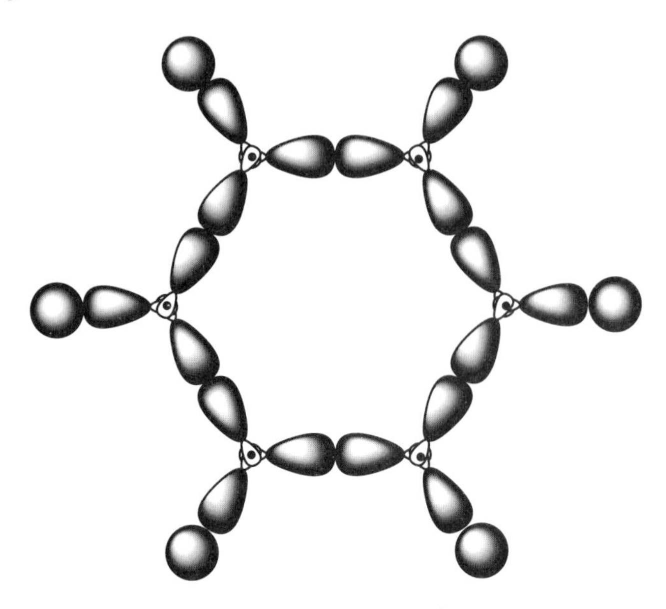

Fig. 5.16 The σ bond framework formed by six sp^2 hybridized carbon atoms in benzene.

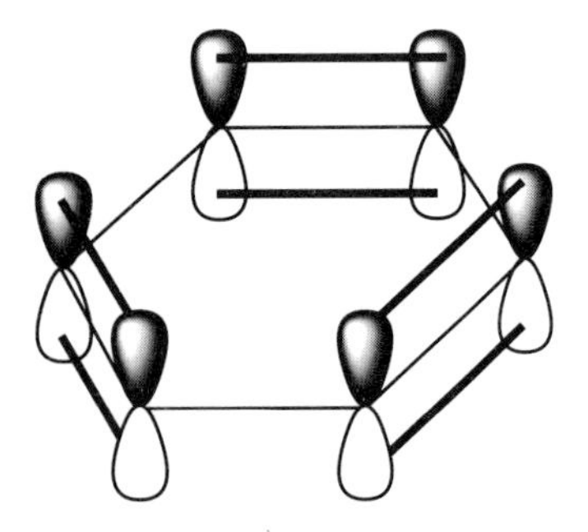

Fig. 5.17 The cyclohexatriene view of benzene

The sum of the total number of C—H, C—C, and C=C bond energies in benzene as modelled by cyclohexatriene is actually short of the experimental value, as determined by the heat of combustion, by about 170 kJ mol^{-1}. The difference is the *resonance stabilization energy* for benzene. Benzene, in the valence orbital approach is regarded as made up from a number of other localized bonding structures, all of which contribute to the overall resonance valence bond model of benzene. These include the two cyclohexatriene structures (Fig. 5.18, often called *Kekulé benzene structures*) and also the three *Dewar benzene structures* which place rather long bonds across the ring. The Dewar structures make a lesser contribution than the Kekulé structures. The real structure of benzene is often denoted with a ring as a shorthand to denote the delocalization.

An alternative method for treating the bonding in molecules such as benzene involves *delocalization* of the π system. Delocalization is more conveniently treated by molecular orbital methods.

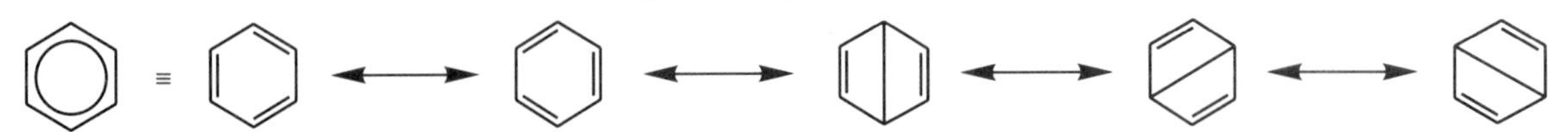

Fig. 5.18 The shorthand representation of benzene (left) is made up from Kekulé and Dewar resonance structures.

5.7 Exercises and problems

1. Describe the bonding in NO_3^-, O_3, H_3O^+, and CO_2 based upon the hybridization model of bonding.
2. What would be an appropriate hybridization description for the central atom in a *square-planar* compound such as $[PtCl_4]^{2-}$? Which orbitals are best suited for this scheme?

6 The MO approach and polyatomic molecules

The concept in which two atomic orbitals are mixed by linear combinations into molecular orbitals (Chapter 3) is not limited to diatomic molecules. For polyatomic molecules, some linear combinations of atomic orbitals delocalize over the *entire* molecule. It is the intention here to introduce some principles pictorially for a few simple molecules. The reader should be aware that the way to determine the form of molecular orbitals for anything other than simple molecules is *not* from a pictorial analysis.

As other books demonstrate, and has already been seen earlier, *symmetry* is extremely important in bonding theory. An understanding of the symmetry of a molecule is approached through a *group theory* analysis. A group theory analysis of the bonding in a molecule allows prediction of which combinations of atomic orbitals lead to bonding, antibonding, and nonbonding molecular orbitals. This is done in a non-pictorial fashion, and once the analysis is complete the results plotted to allow a visualization of the various orbital overlaps.

A group theory analysis does not give a prediction of the shape of a molecule. The VSEPR method is the simplest convenient predictor of molecular shape. Once the shape is assigned, a group theory analysis allows the determination of which linear combination of atomic orbitals are allowed for that shape. Neither does a group theory analysis allow a determination of the energies of the resulting molecular orbitals, only those which are possible. Quantum-mechanical calculations or experimental methods such as photoelectron spectroscopy are necessary in order to give an indication of the energy values of energy levels associated with particular molecular orbitals.

6.1 Beryllium dihydride

An analysis of the molecular orbitals in BeH_2 illustrates some useful principles. A VSEPR analysis predicts a linear structure based upon two electron pairs. As before, the molecular axis is defined as the z-axis.

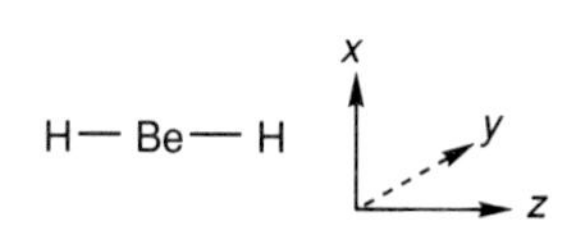

As always in this type of analysis, the aim is not to determine what processes occur on bringing two hydrogen atoms and one beryllium atoms together in some reaction vessel. It is to represent the bonding of a molecule, once formed, using an analysis based upon the atomic orbitals of the component atoms.

It is often convenient to examine first the peripheral atoms, the two hydrogens in this case, and to analyse their mutual interactions. It does not matter that the two hydrogen atoms are quite a distance apart. Since the wave

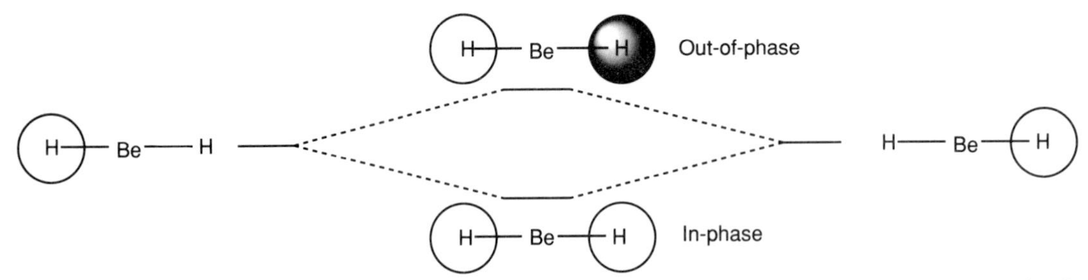

Fig. 6.1 The in-phase and out-of-phase combination of the two hydrogen 1*s* orbitals in BeH_2 as a preparation for interaction with the beryllium atomic orbitals.

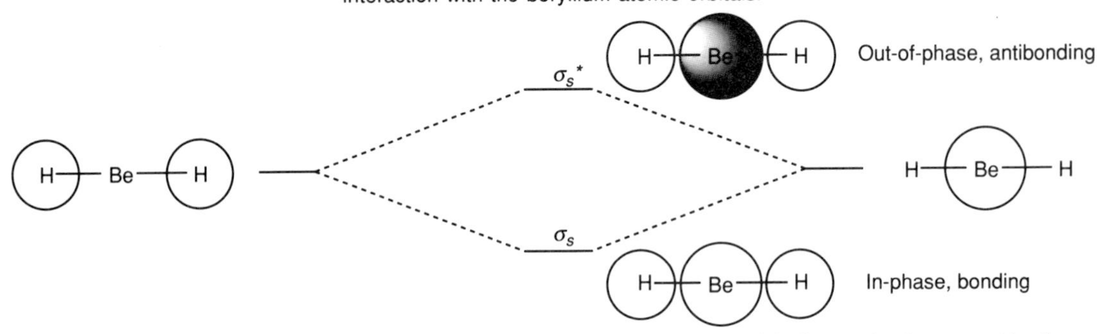

Fig. 6.2 The interaction of the beryllium 2*s* orbital of BeH_2 with the prepared hydrogen in-phase combination.

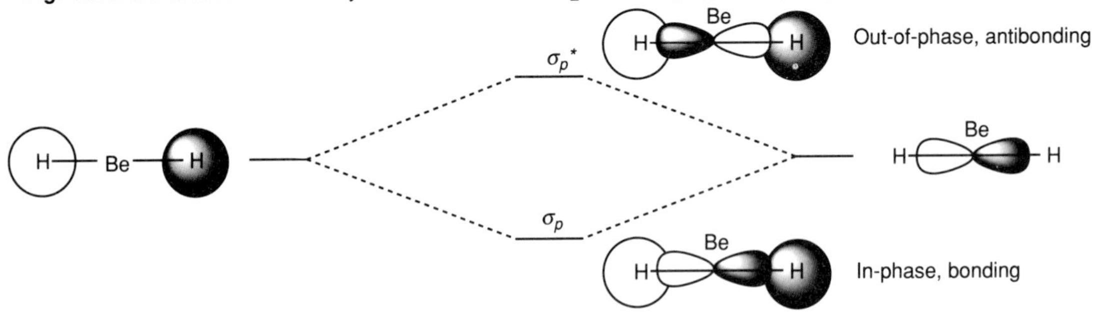

Fig. 6.3 The interaction of the beryllium $2p_z$ orbital of BeH_2 with the prepared hydrogen out-of-phase combination.

functions for the two hydrogen 1*s* orbitals have a value even at the distances involved in BeH_2, there will be at least some interaction.

The discussion of Chapter 3 indicated how two hydrogen 1*s* orbitals interact in H_2. The situation is similar for BeH_2, but the *degree* of interaction is far less because of the greater distance. The result is an in-phase and an out-of-phase combination (Fig. 6.1), with little energy difference between the two. Each of the in-phase and out-of-phase combinations is a *single* orbital consisting of two lobes.

The two hydrogen 1*s* orbitals are now 'prepared' for interaction with the beryllium atomic orbitals. The next step is to determine which beryllium atomic orbitals interact with the prepared in- and out-of-phase hydrogen orbitals. It is clear that the beryllium 2*s* orbital is set to interact with the in-phase combination orbital (Fig. 6.2). On the other hand, the $2p_z$ orbital is set to interact with the out-of-phase combination orbital (Fig. 6.3). In each case the interactions generate new in-phase and out-of-phase combinations. In neither case are the diagrams drawn to scale. In each pair of interactions, the lower orbital is bonding and the upper orbital is antibonding.

This leaves the beryllium $2p_x$ and $2p_y$ orbitals. Inspection shows that there will be zero overlap of these with whatever form the hydrogen orbitals are presented to them. This is illustrated for the $2p_x$ orbital. The beryllium $2p_x$ and $2p_y$ orbitals are therefore unaffected by interaction with hydrogen orbitals and these orbitals *remain* as atomic, non-bonding, orbitals.

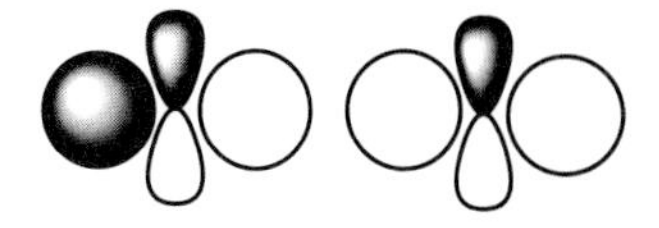

Note that all four orbitals generated through the interaction of the beryllium $2s$ and $2p_x$ orbitals with the hydrogen orbitals possess σ symmetry. The two antibonding combinations are labelled σ^* as before. In addition it is convenient to distinguish the orbitals with the subscript labels s and p to indicate from which beryllium atomic orbitals they originate. Unfortunately it is rather common for textbooks to use a variety of nomenclature for such levels and the reader must therefore be aware of this potential for confusion.

All that remains is to plot the resulting interactions out on an energy level diagram. The pictorial analysis gives no real clue as to the order in which the various molecular orbitals should be written. In fact, it is not normally possible to write down such an order without a quantum-mechanical calculation or spectroscopic measurement, but the ordering in Fig. 6.4 is probably correct.

In the Lewis and hybrid valence bond descriptions of bonding, two equivalent localized Be—H bonds are predicted. In the molecular orbital approach the situation is subtly different. There are four electrons distributed

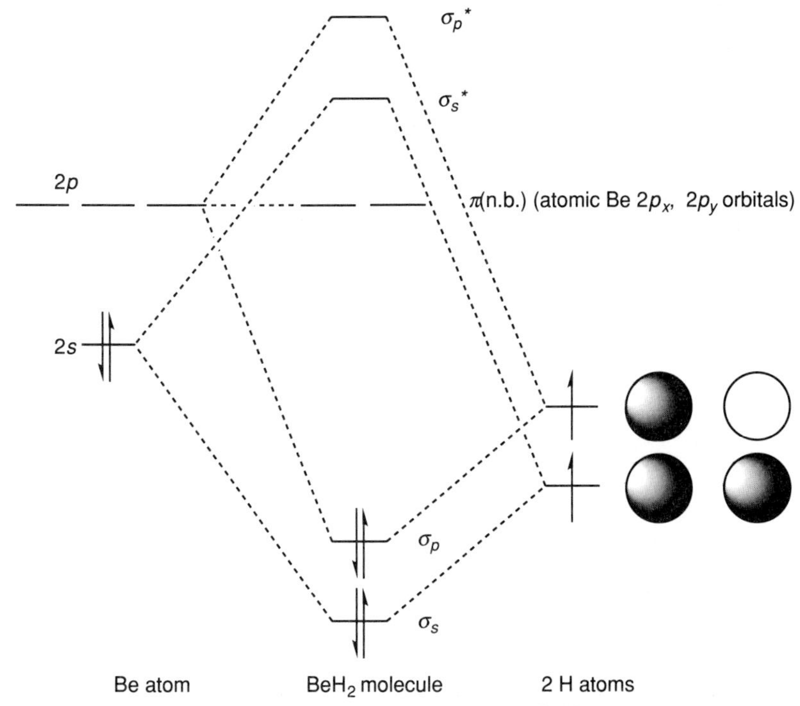

Fig. 6.4 Energy level diagram for BeH_2.

over two bonding σ orbitals, σ_s and σ_p, giving the same total of two bonds. However the bonds are *molecular orbitals* delocalized over *three* atoms and the energy of each molecular orbital is clearly *different*.

6.2 Water

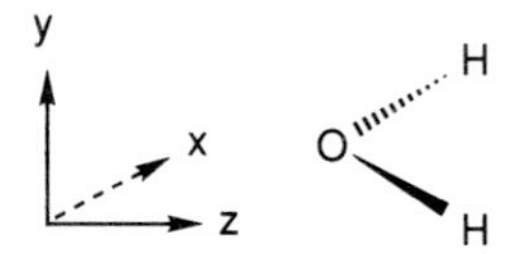

The first step is to define the coordinate axes. By convention and for convenience, the z-axis is drawn through the oxygen atom and along a line between the hydrogen atoms. The plane of the molecule is taken as the xz plane and the plane normal to the molecular plane as the yz plane.

The shape is correctly predicted to be bent by the VSEPR method As for BeH_2, it is convenient to start be examining the interactions of the peripheral hydrogen 1s orbitals (Fig. 6.5). The resulting interactions are closely related to the BeH_2 situation, but the spine of the molecule is now bent.

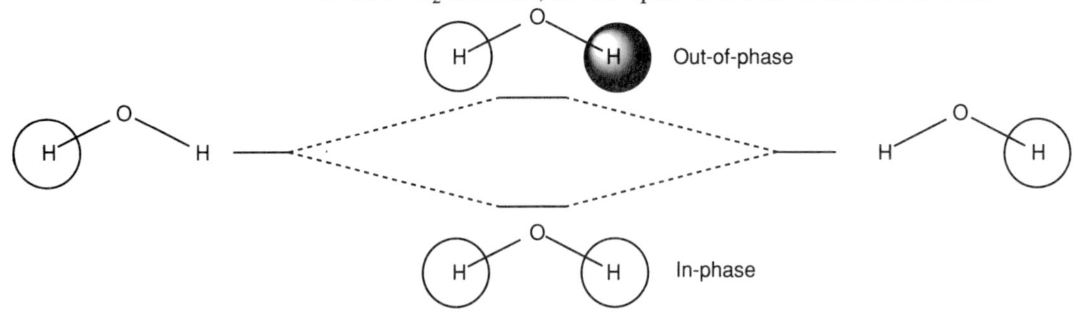

Fig. 6.5 The in-phase and out-of-phase combination of the two hydrogen 1s orbitals in water as a preparation for interaction with the oxygen atomic orbitals.

As for BeH_2 the way forward is to examine which of the oxygen atomic orbitals interact with these two hydrogen combination orbitals. In this case, *two* oxygen orbitals (the 2s and 2p_z orbitals) interact with the in-phase hydrogen combination. These are represented in Fig. 6.6.

This presents a minor complication, and this is how to write down an energy level diagram showing the interaction between *three* (O 2s, O 2p, and 2H 1s in-phase) interacting orbitals. For our purposes it is sufficient to quote the result that if a number, call it n, of orbitals mix together, they do so in such a way as to form exactly the same number, n, of resulting molecular orbitals.

$$n \text{ orbitals} \xrightarrow{\text{orbital mixing}} n \text{ molecular orbitals}$$

In the case of mixing three orbitals, it is generally expected to form one strongly bonding orbital, one strongly antibonding orbital, and one orbital intermediate in nature (Fig. 6.7). This might be non-bonding or weakly bonding or weakly antibonding relative to the component orbitals. Only a

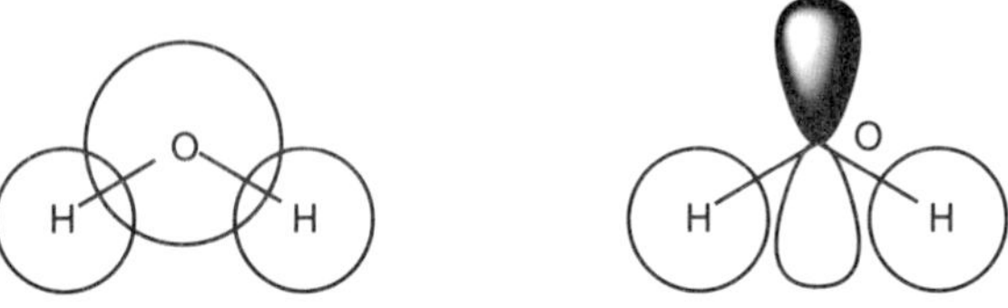

Fig. 6.6 Bonding overlaps between the H 1s orbitals and oxygen 2s (left) and 2p orbitals.

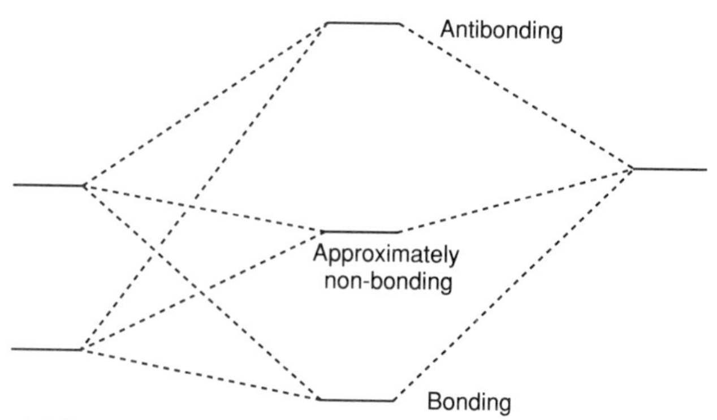

Fig. 6.7 The interaction of two orbitals on one atom with another orbital.

more detailed calculation or measurement is going to give the nature of the intermediate orbital more precisely.

The situation for the oxygen $2p_x$ orbital is simpler and related to the interaction of the beryllium $2p_z$ orbital with the hydrogen out-of-phase combination except that again the spine of the molecule is bent. The resulting bonding and antibonding combination orbitals (Fig. 6.8) are conveniently labelled σ_x and σ^*_x. The remaining oxygen orbital, $2p_y$ is the orbital out of the molecular plane and it is not possible to write down any combination of hydrogen orbitals which display any net overlap with this orbital. The oxygen $2p_y$ orbital thus remains an oxygen atomic orbital in the water molecule and is non-bonding. To a first approximation, it will appear at the same level in the energy level diagram as in the atom.

As for BeH_2, there are two bonding σ-type molecular orbitals. Again their energies are different. The two equivalent electron lone pairs of the Lewis and valence bond models are replaced by two non-equivalent non-bonding molecular orbitals, delocalized in one case over the three atoms (Fig. 6.10).

In addition, the $\sigma_{s,z}$ non-bonding orbital, while labelled non-bonding to make a point, when calculated out is *marginally* bonding and also makes a contribution. This main lobe of this orbital is directed out along the z-axis and is effectively a lone pair, while the minor lobe directed between the

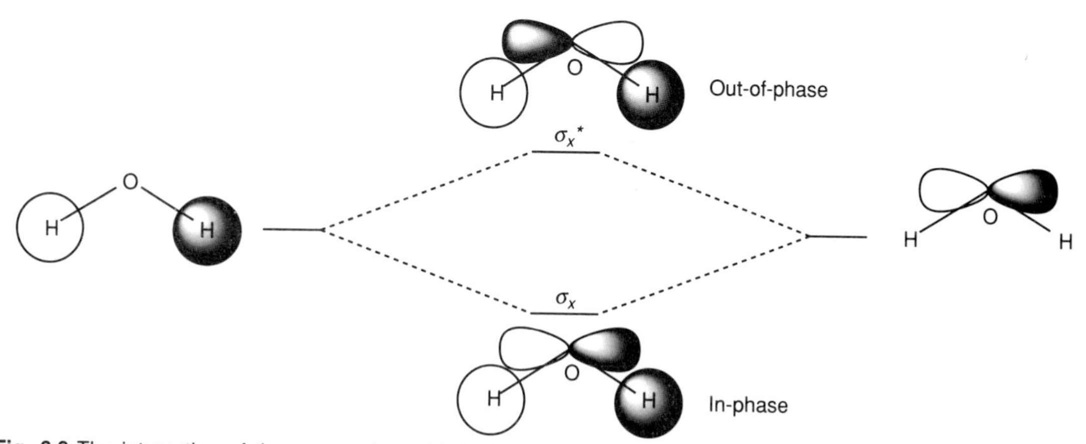

Fig. 6.8 The interaction of the oxygen $2p_x$ orbital of water with the prepared hydrogen out-of-phase combination.

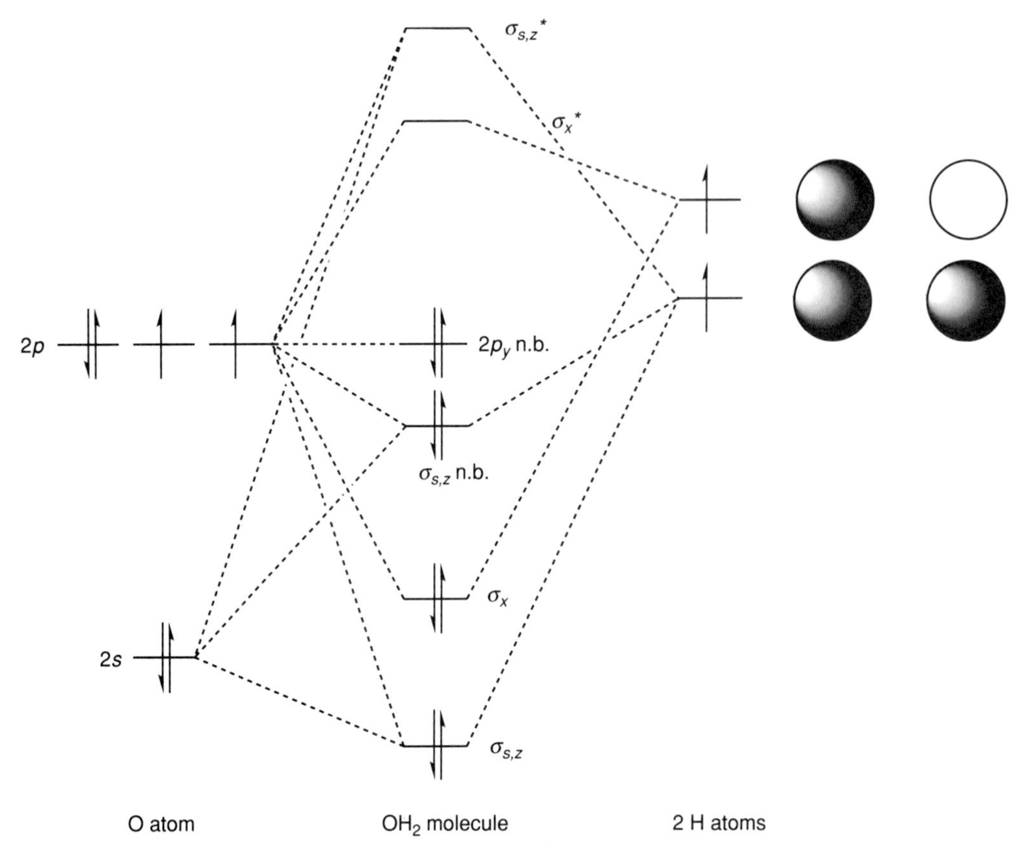

Fig. 6.9 Energy level diagram for water.

hydrogen atoms overlaps with the hydrogen in-phase pair combination to make a weak bonding interaction. The filled atomic $2p_y$ orbital makes the second lone pair. As in the Lewis and valence bond model there are effectively two lone pairs, but their nature is rather different in the molecular orbital model (Fig. 6.10).

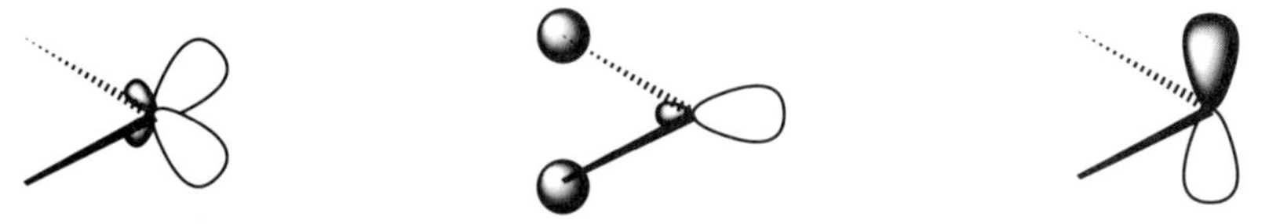

Fig. 6.10 The two lone pairs (left) for water in the valence bond model. The two diagrams to the right represent the two lone pairs in the molecular orbital model.

6.3 The relationship between BeH_2 and H_2O

Both BeH_2 and OH_2 are molecules of an AX_2 type. In effect, BeH_2 is a special case of a general AB_2 molecule whose bond angle is 180°. It is instructive to examine the effect upon the energies of the four occupied molecular orbitals on moving from a linear AX_2 molecule towards a bent molecule with a bond angle of $\theta°$.

The first point to notice is that there is a change of z-axis! By convention, in a linear molecule ($\theta = 180°$) the z-axis is the molecular axis. Also by convention the z-axis in a bent molecule bisects the X—A—X bond. So the instant θ becomes less than 180° the x-axis is relabelled z and *vice versa*. Confusing, perhaps, but that's the way it is.

In a linear molecule the $2p_x$ and $2p_y$ orbitals (which are labelled the $2p_z$ and $2p_y$ orbitals immediately the bond angle is made less than 180°) are non-bonding and degenerate. No matter what the bond angle, the $2p_y$ orbital is non-bonding, so its energy remains constant. However as soon as the bond angle falls below 180° the $2p_z$ orbital ($2p_x$ when the bond angle is 180°) has a net overlap with the in-phase hydrogen 1s pair combination. When the bond angle is very close to 180° it is easy to see that the overlap is very small and that the energy of this bonding orbital must therefore be very close to that of the non-bonding $2p_y$ orbital. The overlap becomes better as the bond angle closes and so its energy drops. When the bond angle is less than 180°, this orbital is labelled σ_z. In water, for which the bond angle ∠ H—O—H = 104.5°, this orbital is slightly bonding. The change of energies with bond angle for all the relevant orbitals can be plotted on a *Walsh diagram* (Fig. 6.11).

There are other orbitals to consider however. Quantum-mechanical calculations also give values for the change in energy of the σ_x (labelled σ_z for $\theta = 180°$) and σ_s orbitals. Significantly, the σ_z orbital ($\theta = 180°$) rises in energy (is destabilized) as θ falls (and is now labelled σ_x). This is because of a decreasing overlap between the $2p_x$ ($\theta < 180°$) orbital and the two hydrogen 1s orbitals as θ falls. On the other hand, the bending process also results in a slight increase in the 1s–1s overlap and this causes σ_s to stabilize very slightly.

The geometry that will be adopted by the real AX_2 molecule will be that in which the *total* energy for *all* the electrons in that molecule is a *minimum*. For the linear BeH_2 only two of the orbitals in question are populated and the minimum is at $\theta = 180°$. For water in which all four orbitals are populated, it is a question of determining the bond angle for which the summed energies of the four populated orbitals is at a minimum. This occurs for $\theta = 104.5°$.

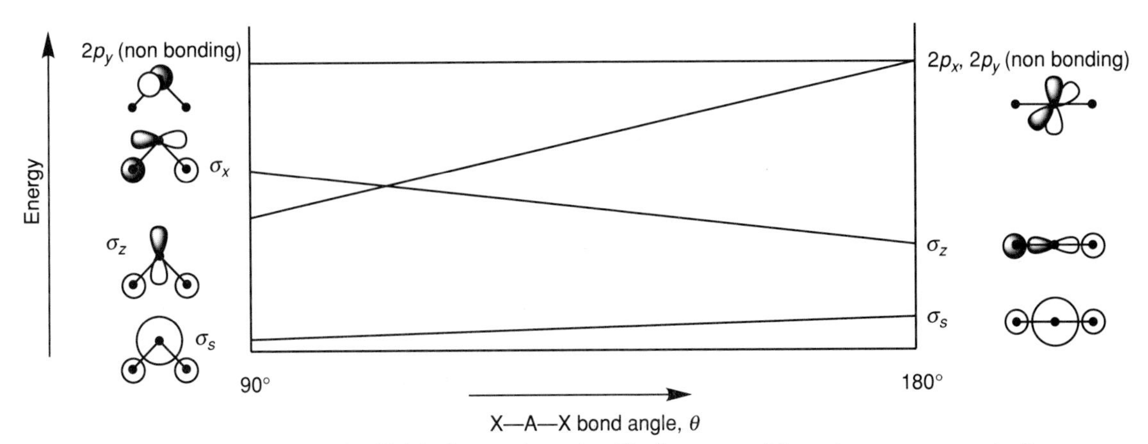

Fig. 6.11 A simplified version of a Walsh diagram for water. The lines are not linear in a more accurate diagram.

6.4 Boron trihydride

Boron trihydride is correctly predicted to be trigonal by the VSEPR method. A simple valence bond representation (Section 5.2) suggests sp^2 hybridization for boron and three equivalent bonds.

The axis definition for BH_3.

The z-axis for BH_3 is taken as out of the plane of the molecule. One way to illustrate the results of a molecular orbital approach is as follows. Again the peripheral hydrogen orbitals are combined together to form a set of orbitals ready to present to the boron atomic orbitals. The situation for the hydrogen orbitals is less obvious than for water since there are three orbitals to mix rather than two. In fact, for a trigonal molecule such as BH_3, these mix into *two* energy levels, the upper of which consists of a degenerate pair of orbitals (Fig. 6.12). The degenerate pair of orbitals look at first sight as though they should not be degenerate, but while formal group theory methods demonstrate that they are, it is beyond the scope of this book to show this.

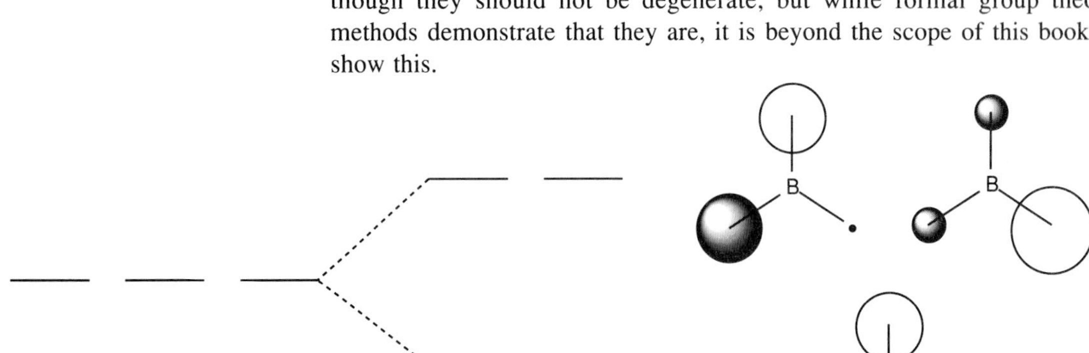

3 H 1*s* orbitals   After preparation for interaction with B

Fig. 6.12 The three combination orbitals arising from mixing the three hydrogen atom 1*s* orbitals in BH_3.

It now remains to consider the mixing of each of these combination orbitals with the boron atomic orbitals. The first point is that the out-of-plane $2p_z$ orbital cannot mix with any of these combination orbitals. The effective overlap is always zero. The boron 2*s* orbital is set to mix with the symmetric hydrogen combination (Fig. 6.13, left). There is a corresponding antibonding combination. The boron $2p_x$ and $2p_y$ orbitals each form bonding interactions with one of the degenerate pair of prepared hydrogen orbitals. In each case, there is a bonding and antibonding combination, although only the bonding combination is illustrated (Fig. 6.13, right).

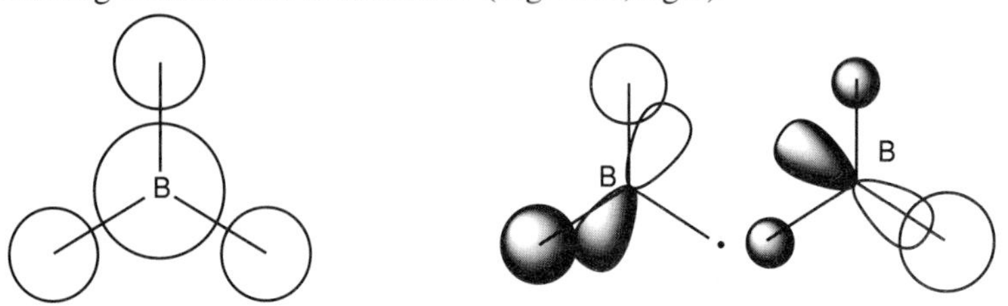

Fig. 6.13 Overlap of hydrogen combination orbitals with atomic Be orbitals.

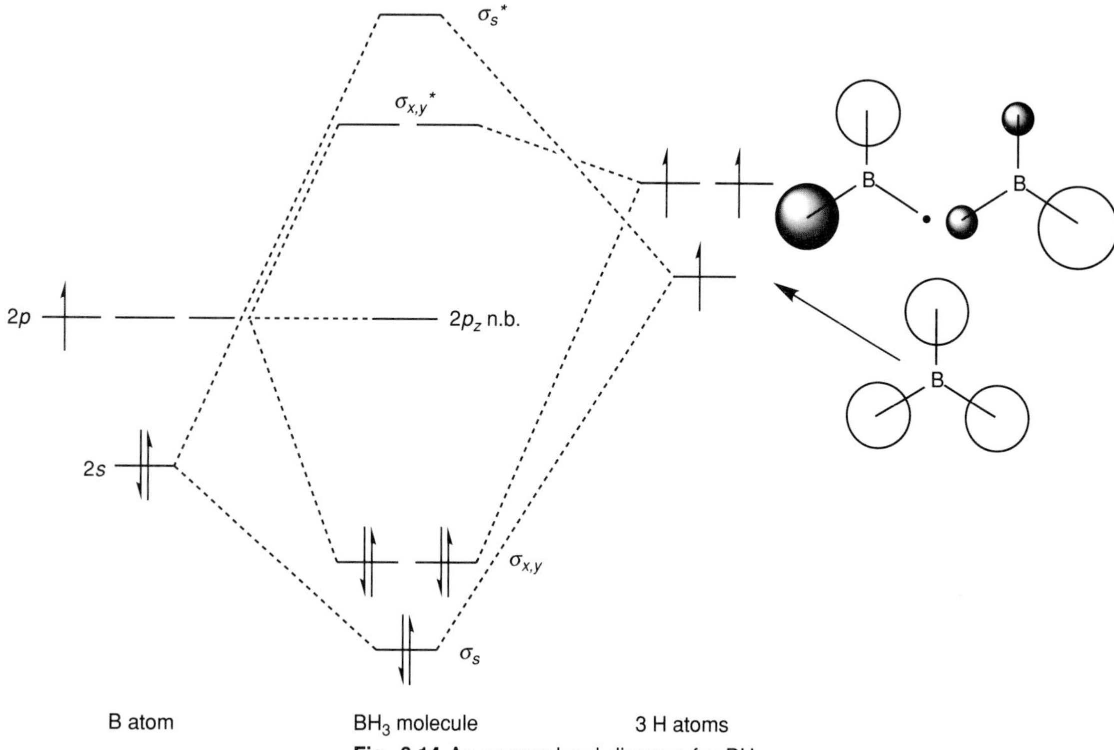

Fig. 6.14 An energy level diagram for BH_3.

The final energy level diagram (Fig. 6.14) demonstrates that there are three bonding B—H σ bonds. Two of the energy levels form a degenerate pair and the third is a little more stable than this pair. Again the molecular orbital representation yields a result subtly different from the Lewis and valence bond representations. The molecular orbital approach gives σ orbitals which are *delocalized* over the whole molecule while the Lewis and valence bond approaches give three equivalent *localized* σ orbitals.

6.5 Boron trifluoride

The situation for BF_3 is related to that of BH_3. In this case the B—F bonds are formed from combinations of fluorine orbitals with the boron $2s$, $2p_x$, and $2p_y$ orbitals and the qualititative appearance of the σ-bonding levels is similar. Our interest centres upon the *filled* three fluorine $2p_z$ orbitals (six electrons total) and the empty boron $2p_z$ orbital. This set of four orbitals overlaps to form a set of four combination molecular orbitals. The lowest of these is the one of interest here since it is the only one containing electrons. It is a symmetric π-type combination orbital (Fig. 6.15). It is occupied by a pair of electrons. It is a bonding orbital since it is lower in energy than all of the four component orbitals.

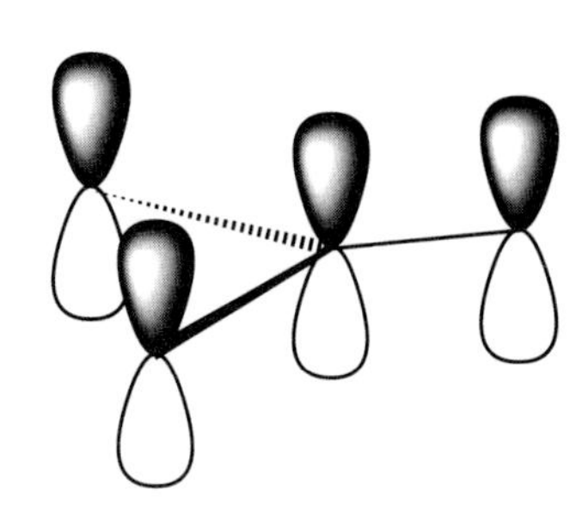

Fig. 6.15 The bonding π orbital in BF_3.

Effectively, some electron density is transferred from what were fluorine lone pairs (residing in $2p_z$ orbitals) to the boron. The boron shares six electrons from the three B—F σ bonds. The filled π-type molecular orbital means that boron has managed to acquire a share of another two electrons, making its effective electron count up to eight, in other words it now *obeys* the octet rule! What about the B—F bond order? It is clear that each B—F interaction consists primarily of a single bond. The extra π interaction is a bond spread between three B—F σ interactions. The effective B—F bond order could therefore be argued to be $1^1/_3$. The multiple B—F bond order is expressed in resonance terms by several resonance structures.

BF_3 (F—B single bonds) $\longleftrightarrow$ F^+=B^- $\longleftrightarrow$ B^-=F^+ $\longleftrightarrow$ ^+F=B^-

6.6 Methane

The approach used for BeH_2 and OH_2 in which the peripheral hydrogen 1*s* orbitals are premixed for presentation to the central atom's atomic orbitals is perfectly feasible for methane, and this is the way the group theory analysis works. However, the results would not be intuitive, and in this case it is better instead to take the central carbon orbitals and see which combinations of peripheral orbitals generate an effective overlap. Essentially, this amounts to the same thing.

This is shown for the carbon 2*s* orbital in Fig. 6.16. If the wave function signs of the peripheral hydrogen 1*s* orbitals are all set to the same as the central 2*s* orbital, then a good bonding combination is formed. The corresponding antibonding combination is drawn simply by reversing the sign of the central atom wave function. The bonding combination is therefore a molecular orbital *delocalized over all five atoms of the methane molecule*.

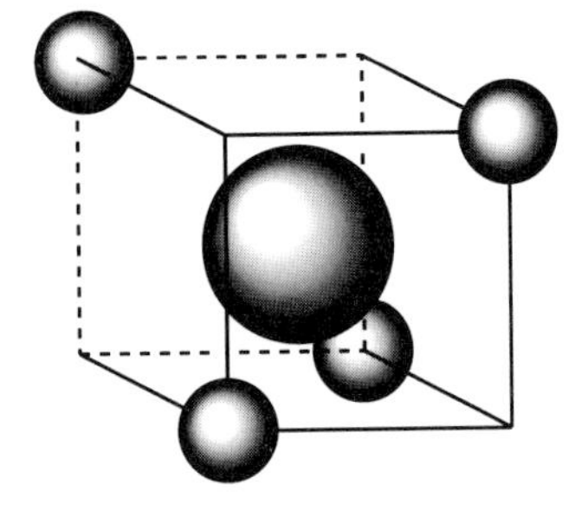

Fig. 6.16 Overlap of the carbon 2*s* orbital with the completely symmetric hydrogen combination orbital.

Consider the carbon 2*p* orbital to the left in Fig. 6.17. A bonding combination is clearly available by writing the wave function sign of the two lower 2*s* orbitals to be the same as the lower lobe of the 2*p* orbital. The same applies for the other two hydrogen 1*s* orbitals and the upper lobe of the 2*p* orbital. The situation for the other two *p* orbitals is precisely similar except in their relative orientations. The three 2*p* orbitals therefore constitute a degenerate group. The antibonding orbitals corresponding to these bonding combinations follows by reversing the signs of the central atom orbitals, and the resulting set of three orbitals is also degenerate.

The resulting energy level diagram is shown in Fig. 6.18. There are four

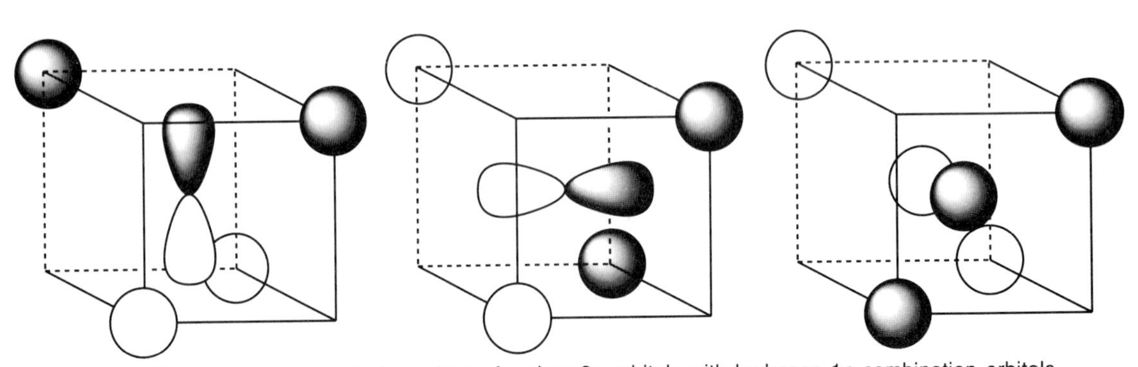

Fig. 6.17 The degenerate set of overlaps of carbon 2*p* orbitals with hydrogen 1*s* combination orbitals.

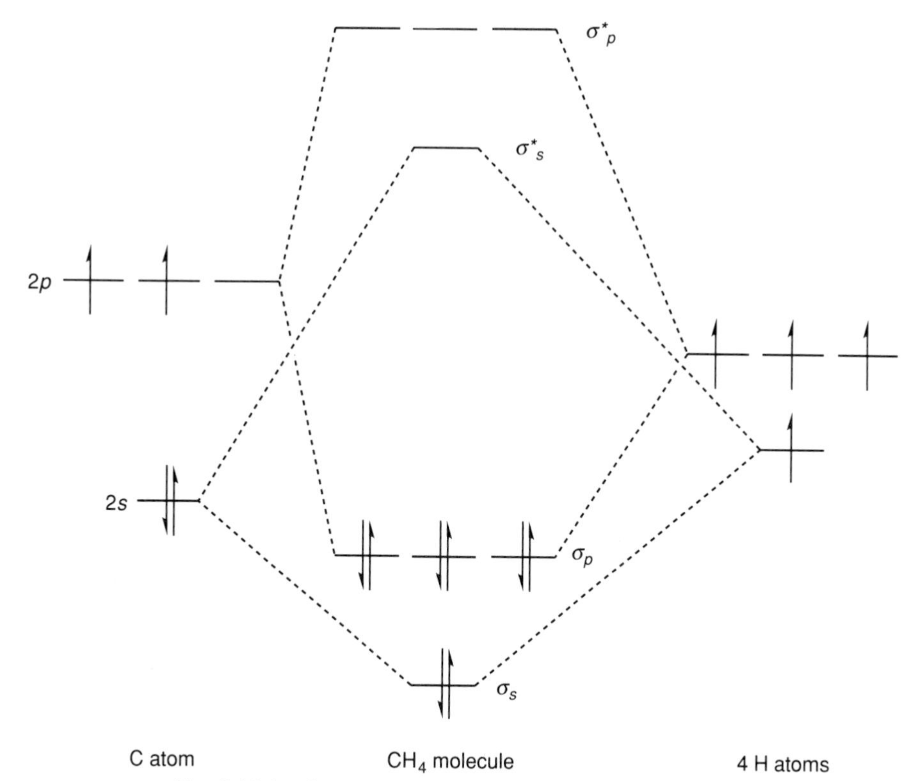

Fig. 6.18 A schematic molecular orbital diagram for methane.

occupied σ molecular orbitals, but the energy of one of these is different from the other three. This is quite distinct from the familiar Lewis and valence orbital picture with its four equivalent bonds constructed from sp^3 hybrids.

Experimental photoelectron spectroscopic data does suggest two distinct bonding levels for methane, and not one as would be expected from the valence bond model. Results such as this are one reason why the molecular orbital model is usually used when considering the spectroscopic properties of molecules.

6.7 Benzene

It is perfectly possible to construct an energy level diagram for benzene, however it is rather complex because each carbon has four valence orbitals and each hydrogen one, making 30 in total and a crowded diagram! For most purposes the σ and σ^* bond framework is relatively uninteresting and it is the interaction of the out-of-plane p_z orbitals with each other that are most of interest from a reactivity point of view.

It is beyond the scope of this book to demonstrate a derivation of the forms of the various linear combinations of the out-of-plane π orbitals, but it is in order to illustrate the result. For benzene, the six atomic $2p$ orbitals combine to form six molecular orbitals. Two pairs of these are degenerate, meaning that there are four energy levels in total. There is a rather interesting

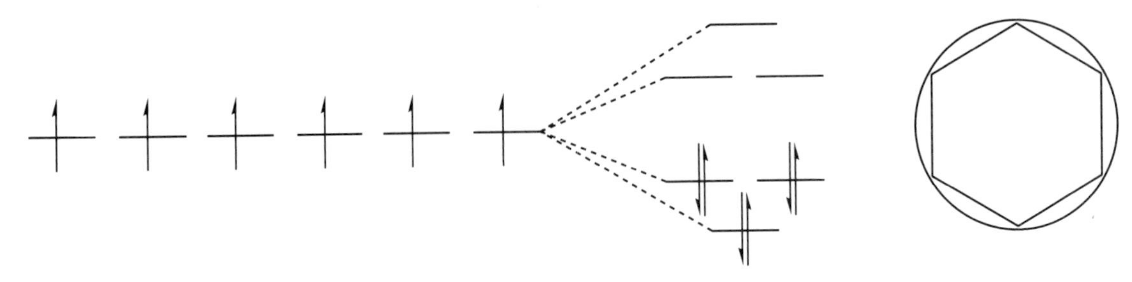

Fig. 6.19 Construction of a diagram to show the relative energy levels for the benzene π orbitals.

geometrical construction that can be used to illustrate (Fig. 6.19) the relative energy of these levels.

If a hexagon is inscribed inside a circle so that the vertices just intersect with the circle, and one vertex is placed at the lowest point of the circle, then the relative height of the other vertices above the lowest represents the relative energy of the other molecular orbital combinations. Each of the carbon $2p_z$ orbitals is populated by one electron. There are therefore six valence electrons in the π system. Inspection of the diagram above shows that these fit neatly into the lower three bonding levels, all of which are bonding. The molecular orbital picture of the bonding therefore involves one very bonding orbital and a degenerate set of two molecular orbitals that are also bonding, but less so. The linear combinations of carbon $2p_z$ orbitals for the bonding orbitals are illustrated in Fig. 6.20.

The most bonding orbital shows no nodal planes other than that in the plane of the molecule. Each of the next lowest orbitals shows one out-of-plane node. While not illustrated, the pattern follows for the antibonding levels with the least antibonding level showing two nodes and the most antibonding showing three nodes.

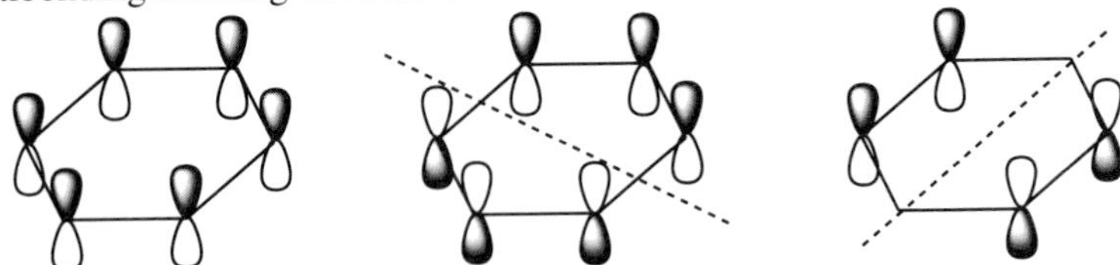

Fig. 6.20 The lowest energy benzene π orbital (left) and the two more weakly bonding benzene π orbitals.

6.8 Exercises

1. Based upon the forms of the *bonding* combinations, write down representations of the antibonding combination orbitals for benzene.
2. Using a similar approach to that suggsted for benzene, write down an energy level diagram to represent the π system in $[C_5H_5]^-$, the cyclopentadienide anion. Sketch the appearance of the molecular orbitals for each of the levels.
3. Describe the bonding in methanal (formaldehyde, $H_2C{=}O$) and ethyne ($HC{\equiv}CH$) based on molecular orbital concepts. Draw energy level diagrams for these molecules.

Reading list

Atkins, P.W., and Beran, J.A. (1992). *General chemistry*, (2nd Edn.). W.H. Freeman & Co., New York, USA.

Bishop, C.B., 1990. Simulation of Rutherford's experiment, *Journal of Chemical Education,* **67**, 889.

Butler, I.S., and Harrod, J.F. (1989). *Inorganic chemistry – principles and applications*, Benjamin/Cummings Publishing Co., Inc., Redwood City, California, USA.

Cotton, F.A., Wilkinson, G., and Gauss, P.L. (1987). *Basic inorganic chemistry*, (2nd edn.). John Wiley and Sons, New York, USA.

DeKock, R.L., and Gray, H.B. (1980). *Chemical structure and bonding*, Benjamin/Cummings Publishing Co., Inc., Menlo Park, California, USA.

Freeman, R.D. (1990), 'New' schemes for applying the Aufbau principle, *Journal of Chemical Education*, **67**, 576.

Gillespie, R.J. 1992. Multiple bonds and the VSEPR model, *Journal of Chemical Education*, **69**, 116.

Gillespie, R.J., (1992). The VSEPR method revisited, *Chemical Society Reviews,* 59.

Gillespie, R.J., and Hargittai, I., (1991). *The VSEPR model of molecular geometry*, Prentice-Hall International, London, England.

Huheey, J.E., Keiter, E.A., and Keiter, R.L. (1993). *Inorganic chemistry – principles of structure and reactivity.* (4th edn.). Harper International., New York, USA.

Jolly, W.L. (1991). *Modern inorganic chemistry*, McGraw–Hill, Inc., New York, USA.

Mackay, K.M. and Mackay, R.A. (1989). *Introduction to modern inorganic chemistry*, (4th edn.). Blackie, London, UK.

Pardo, J.Q. (1989). Teaching a model for writing Lewis structures, *Journal of Chemical Education*, **66**, 456.

Pauling, L. (1992). The nature of the chemical bond – 1992, *Journal of Chemical Education*, **69**, 519.

Porterfield, W.W. (1984). *Inorganic chemistry – A unified approach*, Addison Wesley Publishing Co., Inc., Reading, Massachusetts, USA.

Purcell, K.F., and Kotz, J.C. (1985). *Inorganic chemistry*, (Int. Edn.). Holt Saunders, Japan.

Shriver, D.F., Atkins, P.W., and Langford, C.H. (1990). *Inorganic chemistry*, Oxford University Press, Oxford, England.

Webster, B. (1990). *Chemical bonding theory*, Blackwell Scientific Publications, Oxford, England.

Index